Foundations of
Euclidean and
Non-Euclidean Geometry

PURE AND APPLIED MATHEMATICS

A Program of Monographs, Textbooks, and Lecture Notes

EXECUTIVE EDITORS—MONOGRAPHS, TEXTBOOKS, AND LECTURE NOTES

Earl J. Taft
Rutgers University
New Brunswick, New Jersey

Edwin Hewitt
University of Washington
Seattle, Washington

CHAIRMAN OF THE EDITORIAL BOARD

S. Kobayashi
University of California, Berkeley
Berkeley, California

EDITORIAL BOARD

Glen E. Bredon
Rutgers University

Sigurdur Helgason
Massachusetts Institute of Technology

Marvin Marcus
University of California, Santa Barbara

W. S. Massey
Yale University

Leopoldo Nachbin
Universidade Federal do Rio de Janeiro and University of Rochester

Zuhair Nashed
University of Delaware

Donald Passman
University of Wisconsin

Irving Reiner
University of Illinois at Urbana-Champaign

Fred S. Roberts
Rutgers University

Paul J. Sally, Jr.
University of Chicago

Jane Cronin Scanlon
Rutgers University

Martin Schechter
Yeshiva University

Julius L. Shaneson
Rutgers University

Olga Taussky Todd
California Institute of Technology

MONOGRAPHS AND TEXTBOOKS IN
PURE AND APPLIED MATHEMATICS

1. *K. Yano*, Integral Formulas in Riemannian Geometry (1970) *(out of print)*
2. *S. Kobayashi*, Hyperbolic Manifolds and Holomorphic Mappings (1970) *(out of print)*
3. *V. S. Vladimirov*, Equations of Mathematical Physics (A. Jeffrey, editor; A. Littlewood, translator) (1970) *(out of print)*
4. *B. N. Pshenichnyi*, Necessary Conditions for an Extremum (L. Neustadt, translation editor; K. Makowski, translator) (1971)
5. *L. Narici, E. Beckenstein, and G. Bachman*, Functional Analysis and Valuation Theory (1971)
6. *D. S. Passman*, Infinite Group Rings (1971)
7. *L. Dornhoff*, Group Representation Theory (in two parts). Part A: Ordinary Representation Theory. Part B: Modular Representation Theory (1971, 1972)
8. *W. Boothby and G. L. Weiss (eds.)*, Symmetric Spaces: Short Courses Presented at Washington University (1972)
9. *Y. Matsushima*, Differentiable Manifolds (E. T. Kobayashi, translator) (1972)
10. *L. E. Ward, Jr.*, Topology: An Outline for a First Course (1972) *(out of print)*
11. *A. Babakhanian*, Cohomological Methods in Group Theory (1972)
12. *R. Gilmer*, Multiplicative Ideal Theory (1972)
13. *J. Yeh*, Stochastic Processes and the Wiener Integral (1973) *(out of print)*
14. *J. Barros-Neto*, Introduction to the Theory of Distributions (1973) *(out of print)*
15. *R. Larsen*, Functional Analysis: An Introduction (1973) *(out of print)*
16. *K. Yano and S. Ishihara*, Tangent and Cotangent Bundles: Differential Geometry (1973) *(out of print)*
17. *C. Procesi*, Rings with Polynomial Identities (1973)
18. *R. Hermann*, Geometry, Physics, and Systems (1973)
19. *N. R. Wallach*, Harmonic Analysis on Homogeneous Spaces (1973) *(out of print)*
20. *J. Dieudonné*, Introduction to the Theory of Formal Groups (1973)
21. *I. Vaisman*, Cohomology and Differential Forms (1973)
22. *B.-Y. Chen*, Geometry of Submanifolds (1973)
23. *M. Marcus*, Finite Dimensional Multilinear Algebra (in two parts) (1973, 1975)
24. *R. Larsen*, Banach Algebras: An Introduction (1973)
25. *R. O. Kujala and A. L. Vitter (eds)*, Value Distribution Theory: Part A; Part B. Deficit and Bezout Estimates by Wilhelm Stoll (1973)
26. *K. B. Stolarsky*, Algebraic Numbers and Diophantine Approximation (1974)
27. *A. R. Magid*, The Separable Galois Theory of Commutative Rings (1974)
28. *B. R. McDonald*, Finite Rings with Identity (1974)
29. *J. Satake*, Linear Algebra (S. Koh, T. Akiba, and S. Ihara, translators) (1975)

30. *J. S. Golan*, Localization of Noncommutative Rings (1975)
31. *G. Klambauer*, Mathematical Analysis (1975)
32. *M. K. Agoston*, Algebraic Topology: A First Course (1976)
33. *K. R. Goodearl*, Ring Theory: Nonsingular Rings and Modules (1976)
34. *L. E. Mansfield*, Linear Algebra with Geometric Applications: Selected Topics (1976)
35. *N. J. Pullman*, Matrix Theory and Its Applications (1976)
36. *B. R. McDonald*, Geometric Algebra Over Local Rings (1976)
37. *C. W. Groetsch*, Generalized Inverses of Linear Operators: Representation and Approximation (1977)
38. *J. E. Kuczkowski and J. L. Gersting*, Abstract Algebra: A First Look (1977)
39. *C. O. Christenson and W. L. Voxman*, Aspects of Topology (1977)
40. *M. Nagata*, Field Theory (1977)
41. *R. L. Long*, Algebraic Number Theory (1977)
42. *W. F. Pfeffer*, Integrals and Measures (1977)
43. *R. L. Wheeden and A. Zygmund*, Measure and Integral: An Introduction to Real Analysis (1977)
44. *J. H. Curtiss*, Introduction to Functions of a Complex Variable (1978)
45. *K. Hrbacek and T. Jech*, Introduction to Set Theory (1978) *(out of print)*
46. *W. S. Massey*, Homology and Cohomology Theory (1978)
47. *M. Marcus*, Introduction to Modern Algebra (1978)
48. *E. C. Young*, Vector and Tensor Analysis (1978)
49. *S. B. Nadler, Jr.*, Hyperspaces of Sets (1978)
50. *S. K. Sehgal*, Topics in Group Rings (1978)
51. *A. C. M. van Rooij*, Non-Archimedean Functional Analysis (1978)
52. *L. Corwin and R. Szczarba*, Calculus in Vector Spaces (1979)
53. *C. Sadosky*, Interpolation of Operators and Singular Integrals: An Introduction to Harmonic Analysis (1979)
54. *J. Cronin*, Differential Equations: Introduction and Quantitative Theory (1980)
55. *C. W. Groetsch*, Elements of Applicable Functional Analysis (1980)
56. *I. Vaisman*, Foundations of Three-Dimensional Euclidean Geometry (1980)
57. *H. I. Freedman*, Deterministic Mathematical Models in Population Ecology (1980)
58. *S. B. Chae*, Lebesgue Integration (1980)
59. *C. S. Rees, S. M. Shah, and Č. V. Stanojević*, Theory and Applications of Fourier Analysis (1981)
60. *L. Nachbin*, Introduction to Functional Analysis: Banach Spaces and Differential Calculus (R. M. Aron, translator) (1981)
61. *G. Orzech and M. Orzech*, Plane Algebraic Curves: An Introduction Via Valuations (1981)
62. *R. Johnsonbaugh and W. E. Pfaffenberger*, Foundations of Mathematical Analysis (1981)

63. *W. L. Voxman and R. H. Goetschel*, Advanced Calculus: An Introduction to Modern Analysis (1981)
64. *L. J. Corwin and R. H. Szczarba*, Multivariable Calculus (1982)
65. *V. I. Istrăţescu*, Introduction to Linear Operator Theory (1981)
66. *R. D. Järvinen*, Finite and Infinite Dimensional Linear Spaces: A Comparative Study in Algebraic and Analytic Settings (1981)
67. *J. K. Beem and P. E. Ehrlich*, Global Lorentzian Geometry (1981)
68. *D. L. Armacost*, The Structure of Locally Compact Abelian Groups (1981)
69. *J. W. Brewer and M. K. Smith, eds.*, Emmy Noether: A Tribute to Her and Work (1981)
70. *K. H. Kim*, Boolean Matrix Theory and Applications (1982)
71. *T. W. Wieting*, The Mathematical Theory of Chromatic Plane Ornaments (1982)
72. *D. B. Gauld*, Differential Topology: An Introduction (1982)
73. *R. L. Faber*, Foundations of Euclidean and Non-Euclidean Geometry (1983)

Other Volumes in Preparation

Foundations of Euclidean and Non-Euclidean Geometry

Richard L. Faber

Department of Mathematics
Boston College
Chestnut Hill, Massachusetts

MARCEL DEKKER, INC.　　　　New York and Basel

Library of Congress Cataloging in Publication Data

Faber, Richard L.
  Foundations of Euclidean and non-Euclidean geometry.

  (Monographs and textbooks in pure and applied mathematics ; 73)
   Bibliography: p.
   Includes index.
   1. Geometry. 2. Geometry, Non-Euclidean.
  I. Title. II. Series.
  QA445.F26   1983   516   82-19954
  ISBN 0-8247-1748-1

COPYRIGHT © 1983 by MARCEL DEKKER, INC. ALL RIGHTS RESERVED

Neither this book nor any part may be reproduced or transmitted in any form or by any means, electronic or mechanical, including photocopying, microfilming, and recording, or by any information storage and retrieval system, without permission in writing from the publisher.

MARCEL DEKKER, INC.
270 Madison Avenue, New York, New York 10016

Current printing (last digit):
10 9 8 7 6 5 4 3 2

PRINTED IN THE UNITED STATES OF AMERICA

To Susan and Lynn

PREFACE

Foundations of Euclidean and Non-Euclidean Geometry is intended for use in one- or two-semester courses typically entitled "Foundations of Geometry," "Non-Euclidean Geometry," "Introduction to Modern Geometry," or "Topics in Geometry," and taught to undergraduate mathematics majors, particularly those anticipating a career in secondary school mathematics teaching. It is suitable also for students seeking a Master of Arts in Teaching degree.

The first two chapters (on ancient geometry) are historical, but focus more on the evolution of geometric ideas and their exposition than upon chronology and biographic details. A section is devoted to Greek theories of the universe, viewed as applied geometry. Chapter III describes the axiomatic method, both as employed by the Greeks, and as it is understood today. The incompleteness of Euclid's system of postulates is discussed, and Hilbert's axioms are examined in depth (here and in Appendix B). Chapter IV describes the long but futile quest for a proof of the parallel postulate and the eventual discovery of an alternative geometry. The fundamentals of this geometry are surveyed synthetically in Chapter V, in a manner that blends the approaches of Lobachevsky and Bolyai. Chapter VI derives the trigonometric formulas of hyperbolic geometry from the intrinsic Euclidean geometry of the horosphere. Chapter VII presents a coordinate model for the Lobachevskian plane, while Chapter VIII treats the philosophical impact of the discovery of non-Euclidean geometry and its implications for the study of physical space.

Both the level of difficulty and the prerequisites slowly increase from the beginning to the end of the book. This makes it suitable for a variety of courses taught to different kinds of students:
1. a one-semester course on Geometry from a Historical Perspective (Chapters I–IV and parts of Chapter V);
2. a one-semester course on the Foundations of Geometry (Chapters III–V and Appendix B);
3. a one-semester course on Non-Euclidean Geometry (Chapters III–VI and either VII or VIII, as time permits);
4. a full year course in Geometry (entire text).

Chapters I through V (except for the last few pages of Chapter V, where the exponential function appears) require no background beyond high school geometry. Chapters VI, VII, and VIII require trigonometry and first year calculus. Chapter VII assumes in addition familiarity with the concept of vector. (The relevant facts from vector algebra are summarized in Appendix D.)

Abundant exercises appear at the end of nearly all sections. These are indispensable for the student's progress through the book. Most of the exercises are designed to give the student experience in applying the concepts presented within the text; others call for verification of basic facts used in the text; while some give additional details or alternative explanations of some of the more technical matters. A balance has been struck between computational exercises and theoretical problems.

Richard L. Faber

ACKNOWLEDGMENTS

First and foremost, I would like to thank my lovely wife Susan, who typed much of the manuscript, for abundant patience, understanding, and love. I wish also to thank the following mathematicians and historians of science for their helpful comments and suggestions: Stanley Bezuszka, S. J., Jay P. Fillmore, Thomas Hawkins, Meyer Jordan, John Kenelly, Bennett Kivel, Otto Neugebauer, Kenneth Preskennis, Abraham Sachs, Thomas Tucker, and R. L. Wilder.

I am indebted also to my students for many improvements to the text which resulted from their questions and criticisms.

CONTENTS

PREFACE		v
ACKNOWLEDGMENTS		vii
I	THE BEGINNINGS	1
	1. Mesopotamian Mathematics	2
	2. The Egyptians	25
II	GREEK GEOMETRY	45
	1. Thales of Miletus	46
	2. The Pythagorean School	48
	3. The Athenian School	53
	4. Euclid	59
	5. Archimedes	65
	6. Apollonius	74
	7. Greek Cosmology	80
III	THE AXIOMATIC METHOD	91
	1. Pips and Globs	91
	2. Properties of Axiom Systems	105
	3. Euclid and the Foundations of Geometry	108
IV	HISTORY OF THE PARALLEL POSTULATE	125
	1. The Parallel Postulate	125
	2. Absolute Geometry	130
	3. Absolute Lengths	141

	4.	Saccheri	143
	5.	Lambert	147
	6.	The French Geometers	151
	7.	Wolfgang Bolyai	154
	8.	Gauss	155
	9.	J. Bolyai	160
	10.	Lobachevsky	162
V	**FUNDAMENTALS OF LOBACHEVSKIAN GEOMETRY**		167
	1.	Parallelism of Rays	167
	2.	Angle of Parallelism	169
	3.	Parallelism of Lines—The Angle Criterion	171
	4.	Bisector of a Strip	172
	5.	Properties of $\Pi(x)$	176
	6.	Ideal Points	181
	7.	Ideal Triangles	182
	8.	More Properties of $\Pi(x)$	185
	9.	Divergent Lines	186
	10.	Ultra-Ideal Points	188
	11.	Sheaves of Lines—Fundamental Curves	190
	12.	Limiting Curves	194
	13.	Concentric Horocycles	199
VI	**THE TRIGONOMETRIC FORMULAS**		205
	1.	Perpendicular Lines and Planes	205
	2.	Parallel Lines and Planes	213
	3.	The Limiting Surface	219
	4.	Angle of Parallelism Formula	227
	5.	Triangle Relations	232
	6.	The Three Geometries	239
VII	**THE WEIERSTRASS MODEL**		247
	1.	Preliminaries	248
	2.	H^2	253
	3.	Distance in H^2	257
	4.	Parametric Equation of a Line	262
	5.	Angles	265
	6.	The Homogeneous Representation	269

CONTENTS xi

		7. Parallels and Horocycles	270
		8. Intersections	274
		9. Equidistant Curves	275

VIII LOBACHEVSKIAN GEOMETRY AND PHYSICAL SPACE   279
    1. Defects and the Parallax of Stars   280
    2. The Finite Curved Universe   283
    3. Philosophical Objections:
       Truth or Convenience   288

APPENDIX A: Definitions, Postulates, Propositions of Euclid, Book I   291

APPENDIX B: Hilbert's Postulates   299

APPENDIX C: Hyperbolic Functions   309

APPENDIX D: Vector Geometry and Analysis   313

BIBLIOGRAPHY   321
INDEX   325

Foundations of
Euclidean and
Non-Euclidean Geometry

I

THE BEGINNINGS

Here and elsewhere we shall not obtain the best insight into things until we actually see them growing from the beginning . . .
—Aristotle

I venture to suggest that if one were to ask for that single attribute of the human intellect which would most clearly indicate the degree of civilization of a race, the answer would be, the power of close reasoning, and that this power could best be determined in a general way by the mathematical skill which members of the race displayed. Judged by this standard the Egyptians of the nineteenth century before Christ had a high degree of civilization.
—A. B. Chace

We begin with a survey of earliest recorded geometry, in ancient Mesopotamia (Babylonia) and Egypt. These civilizations did not have anything like the mathematical symbolism which is indispensable to modern mathematicians. They had no symbols for +, -, ×, or ÷, nor did they use letters like x and y to denote unknowns (a practice introduced later by the Greeks). The Egyptians were additionally hampered by a very cumbersome notation for fractions.

Although their mathematics was primitive by contemporary standards—and much of it might not even be called mathematics by some modern critics—some of their achievements were impressive, and doubly so in view of the meager tools at their disposal.

Much of the Egyptian and Babylonian mathematics we shall discuss in this chapter is not geometry at all but arithmetic or algebra, but since geometry was not recognized as a separate science by these cultures, we can gain a fuller understanding of earliest geometry by examining it in the context of ancient mathematics as a whole.

It is important to realize that only during the present century have we amassed any appreciable knowledge of the mathematics of the second and third millenia B.C. Many of the opinions expressed by earlier authors have had to be revised or even reversed because of new discoveries. No doubt many of our present ideas about early mathematics and the cultures in which it was practiced will have to be modified in the future.

1. MESOPOTAMIAN MATHEMATICS

Although it is not possible to designate a single locality as the birthplace of mathematics, it is known that significant mathematical activity was taking place at least as early as the beginning of the third millenium B.C. in the area around the Tigris and Euphrates rivers known as Mesopotamia (literally, "between the rivers") and roughly encompassing present day Iraq. The inhabitants of this region were known as Sumerians, and their civilization took the form of a collection of independent city-states, such as Babylon, Ur, Susa, Nippur, and others. The term Babylonia is often used in reference to the sequence of civilizations that inhabited Mesopotamia, but Babylon was not the center of civilization at most times.

The Sumerians gradually came under the domination of a Semitic people known as the Akkadians, who absorbed much of the Sumerian culture, including their system of numeration. About 1700 B.C., Hammurabi founded the first Babylonian dynasty and established a uniform code of law. Later migrations and invasions brought a series of cultures to power—Assyrians, Chaldeans, Medes, Persians, and Greeks—but there was a continuity in language, science, and mathematics from the Sumerian time down to the time of Alexander the Great, who conquered Mesopotamia around 330 B.C. The last three centuries B.C. were known as the Seleucid period, after one of Alexander's generals (Seleucus), who became regent of the area after Alexander's death.

The Babylonians wrote on clay tablets with a stylus of triangular cross section which produced wedge shaped characters. The clay tablets were then baked in the sun or in a kiln until hard. This form of writing is known as cuneiform, from the Latin *cuneus,* meaning wedge. Literally tens of thousands of these tablets, several hundred of mathematical content, have survived to the present, but as yet only a small fraction of these have been deciphered and catalogued. Figure I-1 illustrates the writing of cuneiform numbers.

Whole numbers below 60 follow a decimal pattern, similar in principle to the numeration of the Egyptians and Romans. However, the symbol for 60 is the same as that for the number 1. In fact, this symbol was used to represent any power of 60, the exponent being determined from context. This means that the Akkadians used a positional notation, or "place value system," in which the value assigned to a digit is determined by its position within the written number. As you know, our own decimal system is positional also. For example, 2864 represents

$$(2 \cdot 10^3) + (8 \cdot 10^2) + (6 \cdot 10^1) + (4 \cdot 10^0)$$

Since the Akkadians used 60 as a base rather than 10, their system was *sexagesimal* rather than decimal. Thus,

$$= (5 \cdot 60^2) + (42 \cdot 60^1) + 31, \text{ or } 20551 \text{ in decimal}$$

Because the early Babylonians had no symbol for zero, their numbers were often ambiguous. The symbol for 20 could represent not only 20, but $20 \cdot 60$, $20 \cdot 60^2$, etc., or even $20/60$ or $20/60^2$, etc., since positional notation was used for fractions also. Moreover, could stand for $60 + 10$, $60^2 + 10$, or $60^3 + 10$, etc., or any of these divided by a positive power of 60. Occasionally a small space was used to signify the absence of a digit in a particular position, and

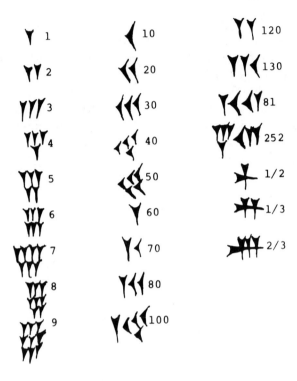

Figure I - 1.

in the Seleucid period a special symbol, ⟨symbol⟩, was used for this purpose, but in interpreting most early Babylonian tablets, one must rely chiefly on context. A few fractions, such as 1/3, 2/3, and 1/2 had special symbols, as shown in Figure I-1.

The use of positional notation was a tremendous aid to calculation, especially in that fractions could be manipulated as easily as whole numbers. (This is a feature of our decimal system also: 0.0163 × 0.21 is computed by exactly the same procedure as is 163 × 21. By contrast, we shall see in the next section that for an Egyptian scribe, the computation of, say, 0.4 × 1.7 was quite different from the simple calculation of 4 × 17.) Another advantage of positional notation is that a small fixed set of symbols suffices for the writing of arbitrarily large numbers. With Roman numerals, for example, one would eventually run out of letters.

Why the Babylonians adopted a sexagesimal system is not entirely clear. It may have resulted from a reconciliation of two different systems of weights or measures, in which the unit of one system was 60 times the unit of the other. Less likely explanations are an early belief that a year consists of 360 days, and the abundance of integral divisors enjoyed by the number 360. Sexagesimal notation was used for writing fractions by the Greek astronomers Hipparchus (2nd century B.C.) and Ptolemy (2nd century A.D.) and in fact throughout Europe down to the 16th century. Our division of hours and degrees into minutes and seconds stems from this practice. (Our present system is somewhat inconsistent though, since we represent hours or degrees decimally, minutes and seconds sexagesimally, and parts of a second decimally: e.g., $117° 31' 14.26''$.)

For convenience, we can transcribe Babylonian numbers by replacing each sexagesimal digit by its decimal equivalent (between 0 and 59). Commas are used to separate the digits and a semicolon denotes the "sexagesimal point," whose location is inferred from context or the nature of the problem at hand. Thus, 2,41;13,20 represents $2 \cdot 60 + 41 + 13/60 + 20/60^2$. 0;30 represents 30/60 or ½.

Nearly all of the tablets that have been deciphered belong either to the time of the Hammurabi dynasty (approx. 1800-1600 B.C., referred to as the "Old-Babylonian" period) or to the Seleucid period (the last three centuries B.C.). Mathematically, the former period is more significant. The mathematical tablets, of which many were apparently school texts intended for apprentice scribes are classified as either table texts (described below) or problem texts (lists of exercise with or without solutions).

The Babylonians were prodigious makers of tables and relied heavily upon these tables for the solution of arithmetic and algebraic problems that arose in commerce, agriculture, and engineering. In addition to tables of products and reciprocals, they compiled tables of squares and square roots, cubes and cube roots, sums of squares and of cubes, exponential functions (used for the computation of compound interest), and what we would today call partial tables of logarithms to the bases 2 and 10.

Table I - 1

2	30	16	3,45	45	1,20
3	20	18	3,20	48	1,15
4	15	20	3	50	1,12
5	12	24	2,30	54	1,6,40
6	10	25	2,24	1	1
8	7,30	27	2,13,20	1,4	56,15
9	6,40	30	2	1,12	50
10	6	32	1,52,30	1,15	48
12	5	36	1,40	1,20	45
15	4	40	1,30	1,21	44,26,40

A typical table of reciprocals is transcribed in Table I-1, reprinted from [32]. The entries are arranged in pairs having a power of 60 as product. By suitable location of the sexagesimal point, this product is unity. Thus,

$$2^{-1} = 0;30 \qquad 40^{-1} = 0;1,30$$
$$3^{-1} = 0;20 \qquad 1,21^{-1} = 0;0,44,26,40$$

Notice that gaps occur in the left hand columns at integers which contain prime factors which do not divide 60. Such numbers, called "irregular," have reciprocals that are nonterminating repeating sexagesimals. For example,

$$7^{-1} = 0;8,34,17,8,34,17,8,34,17,\ldots$$

There are tablets in which the reciprocals of some irregular numbers are approximated.

In place of multiplication tables containing all products $x \cdot y$, for x and y assuming all values from 1 to 59, the Babylonians compiled collections of tables for products $x \cdot y$ in which, in each table, x takes on the values 1 through 19, 20, 30, 40 and 50, while y (fixed) is almost always the reciprocal of a regular number, i.e., a number such as is found in a right hand column of the standard reciprocal table (Table I-1). (A table with y = 7 was often included.) Such a set of tables was used for finding quotients as well as products. For example,

MESOPOTAMIAN MATHEMATICS

$$37 \div 24 = 37 \times 0;2,30 \text{ (from Table I-1)}$$
$$= (30 \times 0;2,30) + (7 \times 0;2,30)$$
$$= 1;15 + 0;17,30 \text{ (multiplication tables)}$$
$$= 1;32,30$$

The Babylonians were not completely bound by these tables, but had developed methods for proceeding in cases to which standard tables were not applicable.

Figure I-2 contains a sketch of an early tablet from the Yale Babylonian collection. The side of the square is labeled 30. Along the diagonal, we read 1;24,51,10. If we multiply these two numbers (which is easily done by dividing the second number by 2, the reciprocal of 0;30), we obtain 42;25,35, which appears just below the diagonal on the tablet. If we convert 1;24,51,10 to decimal, we obtain an excellent approximation of $\sqrt{2}$ (Exercise 13). Neugebauer believes that this approximation was obtained by an iterative method familiar to modern students of analysis: if x_1 is a first approximation to $\sqrt{2}$, then a sequence of better and better approximations x_n is defined recursively by

$$x_{n+1} = \frac{x_n + 2/x_n}{2} \tag{1}$$

Often called Newton's method, this procedure was used by the Greeks also. If we begin with $x_1 = 1;30$, then $x_2 = 1;25$ and $x_3 = 1;24,51,10, \ldots$. Both of these values were used by the Babylonians.

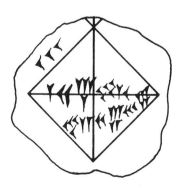

Figure I - 2.

The same technique may have been used in arriving at the approximation, which we would write

$$c \approx a + b^2/(2a) \qquad (2)$$

for the diagonal of a rectangle of sides a and b. If we replace 2 by $a^2 + b^2$ in the numerator of (1), and let $x_1 = a$, then

$$x_2 = \frac{x_1 + (a^2 + b^2)/x_1}{2} = a + \frac{b^2}{(2a)}$$

This approximation is the sum of the first two terms of the binomial expansion

$$(a^2 + b^2)^{1/2} = a(1 + b^2/a^2)^{1/2} \approx a(1 + b^2/2a^2)$$

and is a good estimate when b is small in comparison with a.

Applications of the Pythagorean Theorem abound in all periods of Mesopotamian mathematical history. One Old-Babylonian text presents the following problem and its solution [50, p. 76]:

> A patu (beam?) of length 30 (stands against a wall). The upper end has slipped down a distance 6. How far did the lower end move? Square 30, find 15,0. Take 6 from 30 (find 24). Square 24, find 9,36. Subtract 9,36 from 15,0, find 5,24. 5,24 is how a square? It is 18 squared. 18 along the ground it has moved.

The computation of $[30^2 - (30 - 6)^2]^{1/2}$ is a particular case of the general formula

$$x = [c^2 - (c - d)^2]^{1/2}$$

(see Fig. I-3). The triangle above has sides in the ratio 3:4:5, but different ratios appear in many other tablets. Although mathematical techniques were nearly always given in the form of specific numerical examples, it is clear from the frequent repetition of certain types of problems that general formulas were known.

To the Babylonians, geometry was not a separate science, but rather one among many practical areas to which they could apply their arithmetic and algebraic techniques. Moreover, in the solution of geometric problems, they often summed lengths and areas or

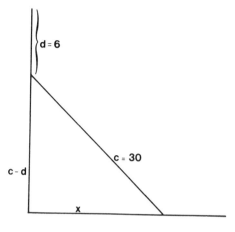

Figure I - 3.

multiplied volumes, even though the results of such operations have no geometric meaning. Similarly, in economic problems, quantities such as wages, numbers of workmen, and man-days of labor are routinely added or multiplied. It is evident from this that words such as length, width, wages, etc., were often treated abstractly, as variables, whose algebraic relationships were the main feature of interest.

Here is another problem, dealing with a volume of earth in the shape of a rectangular box. We are given the area of the base, the volume, and the sum of the length and width of the base, and are asked to find the length, width, and height of the box. (The words SAR, GAR, and kùš refer to units of measure whose exact meaning need not concern us here.)

> ... 7½ SAR the area, 45 SAR the volume; I added the length and the width, and (the result is) 6;30 (GAR). [What are] the length, the width, and [its depth]?

In this passage, taken from [33, p. 72], ½ in place of ;30 indicates that the scribe used the special symbol for ½. Words or numbers in brackets have been restored in translation from illegible characters on the tablet. Words in parentheses are explanatory and were added by Neugebauer and Sachs.

If we denote the length, width, and height by x, y, and z, respectively, then the problem asks us to solve the system

$$xy = 7\tfrac{1}{2}, \quad xyz = 45, \quad x + y = 6;30$$

z is found immediately by dividing the first two equations, and so the problem concerns a system of equations of the form

$$xy = a, \quad x + y = 2b \tag{3}$$

where a and b are given constants. This form of algebraic problem was very common and many other types of quadratic were solved after first being reduced to this type (see Exercise 15).

Although the Babylonian scribes were familiar with the method of elimination and substitution (i.e., substitute $y = a/x$ in the second equation), they preferred a technique in which the unknowns are set equal to half their sum, plus or minus a correction term:

$$x = b + h, \quad y = b - h$$

The first equation then gives

$$a = xy = (b + h)(b - h) = b^2 - h^2$$

(an algebraic relation well known to the Babylonians), from which h could be determined as

$$h = (b^2 - a)^{\tfrac{1}{2}}.$$

Thus,

$$x = b + (b^2 - a)^{\tfrac{1}{2}}, \quad y = b - (b^2 - a)^{\tfrac{1}{2}}$$

The solution given by the scribe does indeed follow this procedure, with a = 7;30, 2b = 6;30 [33, p. 72].

> When you perform (the operations) take the reciprocal of 7 ½, the area, multiply by 45 SAR, the volume, (and) [you] will get 6 (kùš), its depth. Take the reciprocal of its depth, (and) you will get 0;10; [multiply] 0;10 by 45, the volume, (and) you will get 7;30. Hal[ve] the length and the width which I added together, (and) you will get 3;15; multiply [together] 3;15 times 3;15, (and) you will get 10;33,45; take away 7;30 fr[om 10;33,45], (and) you will get 3;3,45; [take its square root, (and)] you will get 1;45; add [1;4]5 to the one, [subtract 1;45] from [the other, (and)] you will get the length and the width. 5 GAR [is the length; 1 ½ GAR is the width.]

In some examples, equations were transformed to simpler form by the introduction of auxiliary variables (Exercise 15). In others, second degree equations were solved essentially by the quadratic formula, perhaps obtained by "completing the square." Since the equations considered arose out of practical problems, there was always at least one positive real root. Negative roots were never considered. Cubic equations and simultaneous non-linear equations in two or more unknowns were also solved. These feats become more remarkable when we reflect that the Babylonians did not have the benefit of our modern algebraic symbolism.

In their geometric problems, the Babylonians often relied upon approximate solutions as in formula (2). The area A of a circle was frequently given as 0;5 (= 1/12) times the square of the circumference, C. By equating this product to the modern expression for the area,

$$A = \pi r^2 = \pi(C/2\pi)^2 = C^2/4\pi$$

we find the approximation $\pi \approx 3$, an estimate that appears also in the Bible, in I Kings, 7, 23 and again in II Chronicles, 4, 2:

> "And he made the molten sea of ten cubits from brim to brim, round in compass ... and a line of thirty cubits did compass it round about."

The passage appears in the description of Solomon's temple. The estimate $\pi \approx 3$ appears also in the Babylonian calculation of the base of a circular segment of given height.

In a tablet discovered at Susa by French archaeologists in 1936, the approximation $\pi \approx 3;7,30 = 3\frac{1}{8}$ is implicit in the calculation of the circumference of a regular hexagon inscribed in a circle ($\pi \approx 3$ obviously will not do here—why?).

The area A of a quadrilateral with consecutive sides a, b, c, and d was often computed in accordance with the rule

$$A = \frac{a+c}{2} \cdot \frac{b+d}{2}$$

the average of one pair of opposite sides times the average of the other pair. This formula yields too large a result, except in the case of a rectangle (see Exercise 4).

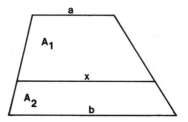

Figure I - 4.

A typical problem arising from the partitioning of land asks for the location and length x of a line which is parallel to the bases of a trapezoid and which divides the trapezoid into two smaller trapezoids of equal areas A_1 and A_2 (Fig. I-4). Using their knowledge of similar figures and proprtion, and the quadrilateral area formula above, the Babylonians were apparently able to discover the correct relation

$$x^2 = (a^2 + b^2)/2$$

(Although the area formula above is incorrect, it does lead to the correct expression for the ratio A_1/A_2 needed to derive the preceding equation.) Unfortunately, we are in the dark as to exactly how this relation was first discovered, since, as in nearly all mathematical tablets, the scribe tells us what the pertinent formulas are by describing a specific numerical example, but he does not reveal the derivation of the formulas.

It was occasionally required to find the volume of a pile of bricks or of a hole in the shape of a truncated square pyramid. If the bases are squares of sides a and b and the altitude is h, then the volume was approximated according to the (incorrect) formula

$$V = \tfrac{1}{2}(a^2 + b^2)h$$

Although the majority of tablets deal with techniques applicable to practical problems, a remarkable Old-Babylonian text, Plimpton 322 (belonging to Columbia University's Plimpton Collection), is number theoretic in character. This tablet deals with right triangles (Fig. I-5) whose sides are whole numbers, i.e., with integral solutions to the Pythagorean equation

MESOPOTAMIAN MATHEMATICS

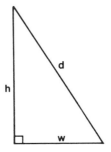

Figure I - 5.

$$d^2 = h^2 + w^2 \qquad (4)$$

Such solutions are called Pythagorean triplets. A left hand portion of the tablet was apparently lost after excavation, so that Plimpton 322 contains only the first four columns in the transcription of Table I-2 (where missing numerals have been restored and certain scribal errors have been corrected in the translation). See [32], [33], [50].

Of these four columns, the middle two were labeled with phrases meaning "width" and "diagonal." They are labeled w and d in the table. The values of h, the height (or longest leg), are not on the preserved tablet but have been computed from the Pythagorean theorem and appear in the fifth column. These values may have been given in one of the missing columns. In all cases, h turns out to be an integer whose prime factors are divisors of 60. As we know, such "regular" numbers have reciprocals with terminating sexagesimal representations and are found in standard Babylonian reciprocal tables (such as Table I-1).

The first column of Plimpton 322 bears a practically illegible heading but is found to contain the values of d^2/h^2. These decrease quite regularly, and in fact the square roots, d/h, decrease almost linearly with the average difference of consecutive values very close to 0;1. (The actual sizes of the triangles, whose sides are all integers, vary quite irregularly.) The angle between sides h and d varies from slightly less than 45° to slightly less than 32°. The column labeled n in Table I-2 is labeled "its name" on the tablet, and serves merely to number the rows of the table.

The author of Plimpton 322 was apparently interested not only in integral solutions of (4), but also in the ratios d/h, which were

Table I - 2*

Plimpton 322							
d²/h²	w	d	n	h	p	q	d/h
1,59,0,15	1,59	2,49	1	2,0	12	5	1;24,30
1,56,56,58,14,50,6,15	56,7	1,20,25	2	57,36	1,4	27	1;23,46,2,30
1,55,7,41,15,33,45	1,16,41	1,50,49	3	1,20,0	1,15	32	1;23,6,45
1,53,10,29,32,52,16	3,31,49	5,9,1	4	3,45,0	2,5	54	1;22,24,16
1,48,54,1,40	1,5	1,37	5	1,12	9	4	1;20,50
1,47,6,41,40	5,19	8,1	6	6,0	20	9	1;20,10
1,43,11,56,28,26,40	38,11	59,1	7	45,0	54	25	1;18,41,20
1,41,33,45,14,3,45	13,19	20,49	8	16,0	32	15	1;18,3,45
1,38,33,36,36	8,1	12,49	9	10,0	25	12	1;16,54
1,35,10,2,28,27,24,26,40	1,22,41	2,16,1	10	1,48,0	1,21	40	1;15,33,53,20
1,33,45	45	1,15	11	1,0	2†	1†	1;15
1,29,21,54,2,15	27,59	48,49	12	40,0	48	25	1;13,13,30
1,27,0,3,45	2,41	4,49	13	4,0	15	8	1;12,15
1,25,48,51,35,6,40	29,31	53,49	14	45,0	50	27	1;11,45,20
1,23,13,46,40	56	1,46	15	1,30	9	5	1;10,40

*Note that in each line, p and q are relatively prime integers, p > q, and, except for line 15, p and q are of different parity. With the exception of lines 11 and 15, each line of the table corresponds to a right triangle whose integer sides have no common factor.
†In line 11, we have divided w, d, and h by the common factor 15 in order to obtain integral values for p and q.

required to lie in a certain range. How might the values in the table have been obtained?

We have seen that a common technique used by the Babylonians in the solution of simultaneous quadratic equations was to set the unknowns equal to one-half their sum plus or minus a correction factor. The same idea may have been applied in determining integral values for w and d for which $\sqrt{(d^2 - w^2)}$ is integral: i.e., setting

$$d = x + y, \quad w = x - y$$

we find $d^2 - w^2 = 4xy$, and so $\sqrt{(d^2 - w^2)} = 2\sqrt{(xy)}$. The latter will be an integer provided x and y are squares of integers: $x = p^2$, $y = q^2$. Consequently, solutions of (4) are obtained by choosing integers p and q and setting

$$d = p^2 + q^2, \quad w = p^2 - q^2, \quad h = 2pq \tag{5}$$

These are the formulas customarily attributed to the disciples of the Greek mathematician and philosopher Pythagoras fourteen centuries later!

MESOPOTAMIAN MATHEMATICS

The values of p and q (corresponding to the given values of d and h) are listed in Table I-2. It is highly significant that p and q are always "regular numbers," and in fact, with the exception of p = 2,5, all values of p and q appear in the standard reciprocal table (Table I-1). It is for this reason that Neugebauer believes that Plimpton 322 was indeed compiled by means of (5). Using combined reciprocal and multiplication tables, the scribe found values of p/q and q/p for which the ratio

$$d/h = \tfrac{1}{2}(p/q + q/p)$$

was close to the desired value. Use of the above procedure would explain also why the sexagesimal values of h are generally "simpler looking" than those of d and w.

We can only guess at the intended purpose of Plimpton 322. Perhaps it served as a source of examples for devising practical problems. In any event, its discovery was no doubt an exciting surprise for scholars of ancient mathematics. It is tempting to speculate on what further number-theoretic lore was known to the Babylonians. Undoubtedly much information is contained in excavated tablets yet to be deciphered.

The facts we do possess are these. The Mesopotamians had developed an efficient system for representing numbers and calculating. They were able to solve a variety of geometric problems, some by exact methods and others by close approximation. Their algebra was highly developed. In addition to the examples presented here, there are numerous examples of the solution of problems involving cubic and higher degree equations in one unknown, systems of simultaneous equations (some non-linear), and arithmetic and geometric progressions. They were interested in general truths and in method. There are several examples of texts in which an entire sequence of several dozen related algebraic problems all have the same numerical solution. Obviously, such texts were intended primarily to teach mathematical method as well as algebraic skill.

Exercises I-1

1. The Babylonians used multiplication tables, together with reciprocal tables, to perform division as well as multiplication. Often a product had to be calculated as a sum of smaller products. For example,

Times 12 Table

1	12
2	24
3	36
4	48
5	1
6	1,12
7	1,24
8	1,36
9	1,48
10	2
20	4
30	6
40	8
50	10

$$53 \times 12 = (50 \times 12) + (3 \times 12)$$
$$= 10,0 + 36 = 10,36.$$

Using the table on this page and Table I-1, pretend you are a Babylonian scribe and calculate the values of

(a) 14×12, (b) 47×12, (c) $32 \times \frac{1}{5}$, (d) $12 \div 8$.

(Do not convert to decimal; use the tables. Show all work including placement of sexagesimal points.)

2. Nearly all Babylonian reciprocal tables contain only reciprocals of "regular" numbers, i.e., numbers of the form $n = 2^a 3^b 5^c$, in which a, b, c are integers. Since the reciprocal of $n = 2^a 3^b 5^c$ is

$$\bar{n} = (0;30)^a (0;20)^b (0;12)^c,$$

all such tables can be computed from knowledge of the reciprocals of 2, 3, and 5. For example,

$$\bar{9} = \bar{3}^2 = (0;20)^2 = 0;6,40$$
$$\overline{45} = \bar{9} \times \bar{5} = (0;6,40)(0;12) = 0;1,20.$$

Compute the reciprocals of 8, 48, 50, 54, 1,15, and 1,21 in the same manner.

3. By performing long division, find the sexagesimal representations of 1/7 and 1/11 and verify that they are repeating.

4. Both the Babylonians and the Egyptians often computed the area of a convex quadrilateral as the product of the average of one pair of opposite sides times the average of the other pair. Show that this approximation yields too large an answer except for the case of a rectangle. (Hint: the area of a triangle is one-half the product of two of its sides times the sine of the included angle.)

5. In a tablet discovered at Susa in 1936, the perimeter of a regular hexagon is computed as 0;57,36 times the circumference of the circumscribing circle. What approximation for π is being used? (The reciprocal of $0;57,36 = 60^{-2} \times 2^7 3^3$ can be deduced in several ways as a product of entries from Table I-1, or you may convert to decimal.)

6. The *Mishnat ha-Middot,* an early Hebrew geometry (c. 150 A.D.), gives the following recipe for finding the area of a circle (tr. by Solomon Gandz, *Quellen und Studien*, Series A, Vol. 2, Belin, 1932):

> "If one wants to measure, let him multiply the thread (diameter) into itself and throw away from it the one seventh and the half of a seventh; the rest is the area, its roof (surface)."

What value is being used here as an approximation to π? (The *Mishnat ha-Middot* reconciles the Biblical $\pi \approx 3$ with this value by claiming that the authors of I, Kings 7, 23 and II, Chronicles 4, 2 are not including the thickness of the container walls in the computation of the circumference. In other words, 30 cubits is the *inside* circumference of a container whose *outside* diameter is given as 10 cubits.)

7. One Old-Babylonian tablet computes the radius of the circumscribed circle of an isosceles triangle with sides 50, 50, and 1,0. Show that the radius is 31;15, as given on the tablet.

8. Letting d = 5,9,1 and h = 3,45,0, compute d/h with the help of the standard reciprocal table (Table I-1) and so verify the correctness of the fourth entry in the last column of Table I-2.

9. In column one of Plimpton 322, opposite n = 8 in Table I-2, the scribe entered 1,41,33,59,3,45 instead of the correct value, 1,41,33,45,14,3,45. In a brief communication in the journal *Centaurus* (18), 1974, C. Anagnostakis and B. Goldstein attempt to explain this error by suggesting that the scribe used the following procedure to compute values of d^2/h^2:

a. compute α = p/q from p and q
b. find ā = 1/α from an expanded table of reciprocals
c. find α² and ā² from a table of squares
d. calculate d²/h² = (α² + ā²)/4 + 0;30

For the entry in question, p = 32 and q = 15.

a. Verify the following

α = 2;8   α² = 4;33,4
ā = 0;28,7,30 ā² = 0;13,11,0,56,15

b. Show that if one neglects the medial zero in ā², then the above procedure will produce the erroneous result found on the original tablet

The following three problems are based upon translations and commentaries in *Mathematical Cuneiform Texts,* by O. Neugebauer and A. Sachs, American Oriental Society, New Haven, Conn., 1945.

10. In the Babylonian tablet MLC 1950 (in the Morgan Library Collection at Yale University), a triangle, which appears to be isosceles, is divided by a line parallel to its base into a smaller isosceles triangle and a trapezoid (figure on this page). We are given the lengths ℓ = 20 and ℓ' = 30 into which the equal sides of the original triangle are partitioned, as well as the area, A = 5,20, of the trapezoid. The scribe asks for the lengths of w_u and $w_ℓ$, called "upper width" and "lower width" in the text.

In typical Babylonian fashion, he solves the problem by first computing $\frac{1}{2}(w_u + w_ℓ)$ and $\frac{1}{2}(w_u - w_ℓ)$. The scribe's computation assumes the approximate formula

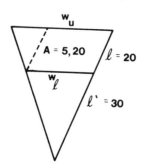

MESOPOTAMIAN MATHEMATICS

$$A = \tfrac{1}{2}(w_u + w_\ell)\ell$$

for the area of the trapezoid. (Of course, this would be exact if the scribe's sketch—nearly obliterated on the tablet—was intended to represent a right triangle with altitude 50.) $\tfrac{1}{2}(w_u + w_\ell)$ is easily found from this area formula.

With the aid of the auxiliary dotted line in the figure, we have, from similar triangles

$$\frac{w_u}{\ell + \ell'} = \frac{w_\ell}{\ell'} = \frac{w_u - w_\ell}{\ell}$$

or

$$w_u = \frac{\ell + \ell'}{\ell}(w_u - w_\ell), \quad w_\ell = \frac{\ell'}{\ell}(w_u - w_\ell)$$

from which we obtain, by adding,

$$w_u + w_\ell = \frac{2\ell' + \ell}{\ell}(w_u - w_\ell)$$

a. From the last equation, deduce that

$$\tfrac{1}{2}(w_u - w_\ell) = \frac{A}{2\ell' + \ell}$$

b. By citing the relevant algebraic equations, explain each of the steps in the scribe's solution [33, p. 48]:

[When you] perform (the operations), take the reciprocal of 20, (and) you will see 0;3. Multiply 0;3 by 5,20, and (the result is) 1[6] .... 30, the length, multiply by 2, add (the resulting) 1,0 and 20, the upper perpendicular, (and the result is) 1,20. Take the reciprocal of 1,20, [multiply] (the resulting) 0;0,45 by 5,20, the area, [and (the result is) 4]. Add [4 to] 16, subtract from 16. [The upper width is] 2[0]; [the lower width is 12.]

11. Tablet YBC 4608 (Yale Babylonian Collection) includes the problem of finding h_1, h_2, b_1, and b_2 in the trapezoid shown here. The data given are

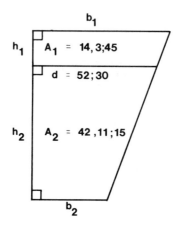

$A_1 = 14,3;45, \quad A_2 = 42,11;15$
$d = 52;30; \quad h_2 = 5h_1$.

The solution, translated below, assumes that the areas are given by formulas

$$A = A_1 + A_2 = (h_1 + h_2)(b_1 + b_2)/2,$$
$$A_1 = h_1(b_1 + d)/2, \quad A_2 = h_2(b_2 + d)/2.$$

These are exact if we assume (as we shall) that the angles on the left in the figure are right, so that $h_1 + h_2$ is the altitude. (The tablet contains no figure.)

In modern terms, the scribe appears to begin his solution by setting $\gamma h_1 = 1$, $\gamma h_2 = 5$, and so $\gamma(h_1 + h_2) = 6$, where γ is a factor to be determined. The next few steps involve the relations

$$\frac{2A}{\gamma(h_1 + h_2)} = \frac{b_1 + b_2}{\gamma}$$

$$\frac{2A_1}{\gamma h_1} = \frac{b_1 + d}{\gamma}, \quad \frac{A_2}{\gamma h_2} = \frac{b_2 + d}{2\gamma}$$

which follow easily from the area formulas.

For each step of the scribe's solution, give the corresponding algebraic equation [33, pp. 50-51].

When you pe[rform (the operations)], let 5 be put (aside) in the lower width*, (and) let 1 be [put (aside)] in the upper width. Add [1] 4,3;45 and 42,11;15, and you will get 56,15. Add 5 and 1, and (the result is) 6; take the reciprocal of 6, and you will get 0;10; multiply [0;10] by 56,15, and multiply the (resulting) 9,22;30 by two: (the result is) 18,45; keep [1] 8,45 in your head. Take the reciprocal of 1 of the upper width, and (the result is) 1; multiply 1 by 14,3;45, and you will get 14,3;45; multiply by 2, (and) you will get 28,7;30. Subtract 18,4[5 fr]o[m 28],7;30, and k[eep] the (resulting) 9,22;30 in your head. Take the [reciprocal] of 5 (of the length) of the lower strip, and (the result is) 0;12; multiply 0;12 by 42,11;15, and you will get [8],26;15. Halve 9,22;30 which you are keeping in your kead, and you will get 4,41;15; add 4,41;15 to 8,26;15, and (the result is) 13,7;30. The reciprocal of 13,7;30 is not obtainable. What should I put to 13,7;30 which will give me 52;30, its dividing-line? [Put] 0;4. The [rec]iprocal of 0;4 will give you 15; you shall multiply 15 by 1, which I put (aside) in the upper width, and the (resulting) 15 is the length of the upper strip. You shall multiply 15 by the 5 which I put (aside) in the lower width, and you will get 1,15, the length of the lower strip. In order for you to see the upper width and the lo[wer] width, multiply the area by two, (and) you will get 28,7;30. Take the reciprocal of 15, and you will get 0;4; multiply 0;4 by 28,7;30, and you will get 1,52;30; tak[e away] 52;30, its dividing-line, from 1,52;30, (and) the (resulting) 1,0 is the upper width. Multiply the whole area by two, (and) [you will get] 1,52,30. Take the reciprocal of 1,30, and you will get 0;0,40; multiply 0;0,40 by 1,52,30, and you will get 1,15; take away 1,0, the upper width, from 1,15, (and) the (resulting) 15 is the lower width.

12. A problem on the tablet YBC 7164 deals with the digging of a canal with rectangular cross section and dimensions

length = 5,0 GAR
width = 3 cubits
depth = 3 cubits

The scribe assumes there are two layers of earth of different densities, so that the rate of digging for the top layer is different from that for the bottom layer. Thus, the depth is divided into two parts, $d' = 1$ cubit (top layer) and $d'' = 2$ cubits (bottom layer). The corresponding "assignments" (i.e., daily digging rates) are given as $a' = 0;20$ SAR and $a'' = 0;10$ SAR. This means that if a worker were to spend a full day in digging through earth from the top layer, he would remove 0;20 SAR of dirt; and analogously for the lower layer.

In actuality, a worker would spend only a certain portion of a day, t', in the upper layer, and the remaining portion, t'', in the lower layer. Since each of these time intervals is directly proportional to

---
*Throughout this problem, the scribe used, for both "length" and "width," a word whose literal meaning is "width."

the relevant layer's depth and inversely proportional to its digging rate, we have

$$t'/t'' = (d'/a') \div (d''/a'')$$
$$t' + t'' = 1$$
$$a = t'a' + t''a''$$

where a is the total assignment, i.e., the total amount of earth removed by a worker in one day.

Find the values of t', t'', and a. (The same tablet contains a similar problem involving three layers with different digging rates. Unfortunately, for both problems the correct answers are given, but no account of the method employed.)

13. With the aid of a pocket electronic calculator, you can easily convert a sexagesimal number less than 1,0 to decimal by a sequence of additions and divisions by 60. For example,

$$1;33,18,43,12 = \frac{\frac{\frac{\frac{12}{60} + 43}{60} + 18}{60} + 33}{60} + 1}{60} = 1.5552$$

a. Convert the Babylonian estimate for $\sqrt{2}$, 1;24,51,10, to decimal. (It differs from the exact value, 1.41421356..., by less than 6×10^{-7}.)

b. Convert 1;22,24,16 and 4;15,33,53,20 to decimal.

A sexagesimal integer is easily converted to decimal by a sequence of additions and multiplications by 60. For example,

$$43,20,35,5 = [(43 \cdot 60 + 20) \cdot 60 + 35] \cdot 60 + 5$$
$$= 9362105$$

c. Convert 9,22,30 and 4,36,28,48 to decimal integers.

14. A decimal integer can be converted to sexagesimal by repeated division by 60. In the following example, the successive remainders, read from bottom to top, constitute the sexagesimal representation of decimal 9362105.

$$9362105 = 156035 \times 60 + 5$$
$$156035 = 2600 \times 60 + 35$$
$$2600 = 43 \times 60 + 20$$
$$43 = 0 \times 60 + 43$$
$$9362105 = 43,20,35,5$$

For those with access to an electronic calculator, the following procedure, applicable to non-integral as well as integral data, can be performed quite rapidly.

First, assume x is a positive number less than 60. Inductively, define

$$x_0 = x, \quad x_{n+1} = 60(x_n - \text{int } x_n)$$

for n = 1,2,3, . . . , where int denotes the "integer part" function. For each $n \geqslant 0$, let $d_n = \text{int } x_n$. Then the sexagesimal representation of x is

$$d_0 \; ; \; d_1, d_2, d_3, d_4, \ldots$$

If x is regular, the process is terminated as soon as an integral x_n is found. (Often in practice, because of calculator round-off, the last x_n will be only very close to an integer.) To apply the method to a number x larger than 60, let r be the smallest positive integer for which $x' = x/60^r < 60$. Apply the above technique to x', and then shift the sexagesimal point r places to the right.

Example: x = 9362105
$\qquad x' = 9362105/60^3 = 43.3430787$
$x_0 = 43.3430787 \quad d_0 = 43$
$x_1 = 20.584722 \quad d_1 = 20$
$x_2 = 35.08332 \quad d_2 = 35$
$x_3 = 4.9992 \quad d_3 = 5$
(Allowing for round-off, x_3 taken as 5)

Therefore,

$x' = 43;20,35,5$, and so (since r = 3)
$x = 43,20,35,5$

Convert to sexagesimal:

(a) 1.4825   (b) 2.06845   (c) 4325.6

15. The tablet AO 8862 (Antiquités Orientales, Louvre, Paris) involves an algebraic problem which was solved through a change of variable [50, p. 63]:

> Length, width. I have multiplied length and width, thus obtaining the area. Then I added to the area, the excess of the length over the width: 3,3 (... the result.) Moreover, I have added length and width: 27. Required length, width, and area.

Letting x = length and y = width, we may state this problem in terms of a system of simultaneous equations:

$$xy + y - y = 3,3, \quad x + y = 27$$

a. By introducing the auxiliary variable $y' = y + 2$, transform the above system into one of the form

$$xy' = a, \quad x + y' = 2b$$

(where a and b are constants).

b. Explain the steps in the scribe's solution:

One follows this method:

27 + 3,3 = 3,30
2 + 27 = 29

Take one half of 29 (this gives 14;30).

14;30 × 14;30 = 3,30;15
3,30;15 - 3,30 = 0;15

The square root of 0;15 is 0;30.

14;30 + 0;30 = 15 length
14;30 - 0;30 = 14 width

MESOPOTAMIAN MATHEMATICS

Subtract 2, which has been added to 27, from 14, the width. 12 is the actual width.

(Translated by B. L. van der Waerden in [50], from O. Neugebauer's translation in *Mathematische Keilschrifttexte,* Quellen und Studien, A 3, Berlin, 1935.)

2. THE EGYPTIANS

The form of writing which the ancient Egyptians carved or painted on stone and wood is known as *hieroglyphics* (literally, "sacred carving," from an early mistaken belief that it was used primarily by the priests). It is a form of picture writing developed during the fourth millenium B.C.

Sometime during the time of the Old Kingdom (3200-2000 B.C.), a cursive or handwritten form known as *hieratic* came into general use. This was written on "paper" that was made by pressing together strips of Nile River papyrus reeds. A single reed, dipped in a type of die, served as a pen. Both hieroglyphics and hieratic were written from right to left. Nearly all surviving documents are in hieratic script. Egyptologists customarily transliterate these into the printed hieroglyphics, and then translate the hieroglyphics into a modern language. Most mathematical texts date from the Middle Kingdom (2000-1800 B.C.).

In order to appreciate the mathematics of the Egyptians, it is necessary first to know something of their methods of arithmetic calculation. The Egyptians employed a decimal (but not positional) system similar in principle to that used by the Romans. Any positive integer was represented as a combination of the following hieroglyphic symbols: | one, ∩ ten, ဌ one hundred, ⚡ one thousand, etc. For example, 327 would appear (written from right to left) as

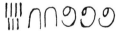

Addition was simply a matter of tallying like symbols and converting a carry into units of the next higher order.

Figure I-6 illustrates some hieratic numbers. In hieratic, addition is less straightforward, and tables may have been used.

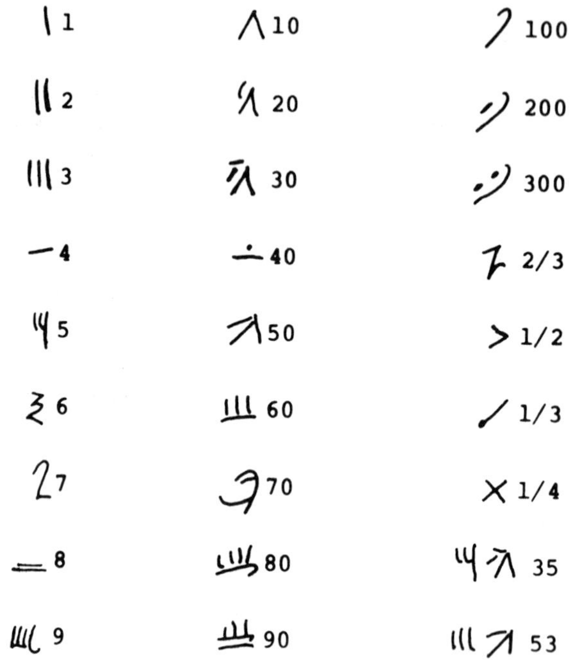

1	∧ 10	⁊ 100
‖ 2	ᛋ 20	⁊⁊ 200
⫼ 3	⩘ 30	⁞⁊ 300
— 4	∸ 40	⁊ 2/3
⫼ 5	⩘ 50	> 1/2
⟨ 6	⫼⫼ 60	⟋ 1/3
⟨ 7	⟋ 70	× 1/4
⩵ 8	⫼⫼ 80	⫼ ⩘ 35
⫼⫼ 9	⫼⫼ 90	⫼⫼ ⩘ 53

Figure I - 6.

Multiplication, which was accomplished through repeated doubling and adding, is best illustrated with examples (where the hieratic has been translated into our Arabic numerals).

```
  (i) multiply 9 by 6      (ii) multiply 21 by 19
        1     9              / 1      21
      / 2    18              / 2      42
      / 4    36                4      84
            ─────              8     168
             54              / 16    336
                                    ─────
                                     399
```

The diagonal marks indicate which multiples of the multiplicand are to be summed. Interestingly, a vestige of the Egyptian multiplication method still survives in present-day digital computers, where numbers are represented internally by sums of powers of two, i.e.,

THE EGYPTIANS 27

in binary notation, and where multiplication is performed electronically by repeated duplication and addition.

A similar process sufficed for division. Thus, to divide 7 into 175, the scribe would "add with 7 until 175":

```
  / 1       7
    2      14
    4      28
  / 8      56
  /16     112
         ___
         175
```

(The quotient is therefore $1 + 8 + 16 = 25$.) Of course, the division might not go evenly, and the scribe would have to resort to fractional multiples of the divisor, which brings us to the Egyptian representation of fractions.

Certain fractions, such as 1/2, 1/3, 2/3, and 1/4, were given special names and symbols. All other fractions of the form 1/n were indicated in hieroglyphics by placing a special symbol, ⌒, meaning "part," above the symbol(s) (or, in some texts, right-most symbol) of the denominator. For example,

$$\text{|||} = 1/5, \quad \text{||∩} = 1/12.$$

In hieratic, a dot was placed over the right-most digit of the denominator. Thus,

$$\text{ⅰⅰ} = 1/5, \quad \text{((∧} = 1/12.$$

In transcribing unit fractions, we shall overscore the numerator with a bar: e.g., $\bar{5} = 1/5$, $\overline{12} = 1/12$, etc. The symbol for 2/3 is customarily transcribed as $\bar{3}$.

Except for 2/3, the Egyptians wrote every non-unit fraction p/q as a sum of distinct unit fractions. For instance, $2/5 = \bar{3}\ \overline{15}$ (meaning the sum of $\bar{3}$ and $\overline{15}$).

According to Van der Waerden, an inkling as to why the Egyptians never devised a more efficient way of representing non-unit fractions can perhaps be found in the literal meaning of their verbal expression for 2/3: "the two parts." Evidently, 1/3 was thought of as "the third part," which, when added to the two parts, gives the whole. The Greeks also referred to 1/3 as "the third part," as for example in Homer's *Iliad*: "Two parts of the night are past, the third part remains." Likewise, 1/4 was called "the fourth part," which completed "the three parts" (3/4) to form the whole. A similar interpretation could be given for "the nth part," for any n. Linguistically then, an expression such as "two-fifths" would be inappropriate, since there is only one fifth part, namely, the last one, which completes the whole. A similar usage appears in the Bible, in Genesis, 47, 24: ". . . ye shall give the fifth part unto Pharaoh, and four parts shall be your own."

Of course, it could still be argued that special symbols or abbreviations could have been invented for sums such as $\bar{5} + \bar{5}$ or $\bar{7} + \bar{7} + \bar{7} + \bar{7}$, etc., but the conservative, tradition-bound Egyptians did not have the modern mathematician's propensity for introducing a new symbol for each frequently occurring expression.

Nevertheless, the Egyptian scribes of the Middle Kingdom (2000-1800 B.C.) had very clever and effective techniques for manipulating unit fractions, so effective in fact that the "art of Egyptian calculation" was taught in Greece, even though the Greeks knew how to work with non-unit fractions by such methods as finding lowest common denominators and reducing to lowest terms.

No doubt every scribe had to memorize certain basic equalities involving the fractions $\bar{2}, \bar{3}, \bar{\bar{3}}$, and $\bar{6}$, such as

$$\bar{6} + \bar{6} = \bar{3} \tag{6a}$$

$$\bar{3} + \bar{6} = \bar{2} \tag{6b}$$

$$\bar{2} + \bar{3} = \bar{\bar{3}} + \bar{6} \tag{6c}$$

$$\bar{2} + \bar{6} = \bar{3} + \bar{3} = \bar{\bar{3}} \tag{6d}$$

and so on. Further rules could be derived from these by multiplying the dominators on both sides by the same factor: e.g., from (6b), $\bar{6} + \overline{12} = \bar{4}, \bar{9} + \overline{18} = \bar{6}$, etc. In addition, the Egyptians, like their Mesopotamian counterparts, relied upon tables.

Since multiplication entailed duplication, a table of the doubles of unit fractions was indispensable. The Rhind Mathematicial Papyrus (RMP [8], written about 1650 B.C. by the scribe A'h-mosè or Ahmes, but (according to its preface) copied from works written about 200 years earlier, begins with a table of twice the unit fractions with odd denominators from 3 to 101.

No doubt many of the entries in this table were obtained as particular cases of general rules of thumb which had been learned centuries earlier from trial and error. For example, for all denominators divisible by 3, the rule $2 \cdot \bar{3n} = \bar{2n} + \bar{6n}$ [cf. Eq. (6d)] was consistently followed. A key step in the original construction of such a table must have been the realization that the duplication of $\bar{n}$ was equivalent to "adding with n until 2 is obtained," i.e., dividing n into 2. That we take this principle as obvious is largely a consequence of our modern notation.

In the calculation of these and other quotients, the sequences

$$\bar{\bar{3}}, \bar{3}, \bar{6}, \bar{12}, \bar{24}, \ldots$$
$$\bar{2}, \bar{4}, \bar{8}, \bar{16}, \bar{32}, \ldots$$

played an important rule, in that the scribe often seems to attempt to reach an answer which is a sum of terms from one or the other of these sequences. For example $2 \cdot \bar{11} = \bar{6}\,\bar{66}$ was calculated as follows (juxtaposition denotes addition):

(What part is 2 of) 11? $\bar{6}$ (of 11 is) $1\,\bar{\bar{3}}\,\bar{6}$. $\bar{66}$ (of 11 is) $\bar{6}$.
(Computation:)

	1	11
	$\bar{\bar{3}}$	$7\,\bar{3}$
	$\bar{3}$	$3\,\bar{\bar{3}}$
/	$\bar{6}$	$1\,\bar{\bar{3}}\,\bar{6}$
/	$\bar{66}$	$\bar{6}$

(Total 2)

The scribe may have first obtained line three of his computation as $3\,\bar{2}\,\bar{6}$ and then applied (6d). Similarly, line four results from applying (6c) to $1\,\bar{2}\,\bar{3}$. Notice that $\bar{3}$ of a number is found by halving $\bar{\bar{3}}$ of it. This was standard practice, and was used even for working out $\bar{3}$ of 3 and even $\bar{3}$ of 1! Most likely, tables of the multiples of $\bar{\bar{3}}$ by both whole numbers and unit fractions were available. The last line of the above example was apparently suggested to the scribe by the need to obtain $\bar{6}$ so as to complete $1\ \bar{3}\ \bar{6}$ to 2.

Similarly, use of the sequence $\bar{2}, \bar{4}, \bar{8}, \ldots$ gave the quotient $2 \div 13 = \bar{8}\ \overline{52}\ \overline{104}$.

$$\begin{array}{rl} 1 & 13 \\ \bar{2} & 6\ \bar{2} \\ \bar{4} & 3\ \bar{4} \\ /\ \bar{8} & 1\ \bar{2}\ \bar{8} \\ /\ \overline{52} & \bar{4} \\ /\ \overline{104} & \bar{8} \end{array}$$

Additional examples appear in the exercises.

The representation of a number as a sum of distinct unit fractions is certainly not unique: e.g., $2/9 = \bar{5}\ \overline{45} = \bar{6}\ \overline{18}$; $2/45 = \overline{30}\ \overline{90} = \overline{36}\ \overline{60} = \overline{27}\ \overline{135}$. Ahmes chose the "simplest" representation, where "simplest" was perhaps determined in accordance with certain guidelines, such as: small denominators are preferred over large; even denominators are preferred over odd (in order to facilitate duplication); the largest simple fraction, $\bar{\bar{3}}$, is preferred, and so on. If you are interested in learning more about Egyptian arithmetic, consult Gillings's *Mathematics in the Time of the Pharaohs* [15].

You may well ask why the Egyptians could not be content with writing simply $\overline{11}\ \overline{11}$ for 2/11 or $\bar{7}\ \bar{7}\ \bar{7}$ for 3/7, etc. However, this would lead to longer and longer strings of unit fractions in the calculation of products by repeated duplication. The Egyptians wanted answers with as few summands as possible.

Following the table of quotients of 2 by the odd integers from 3 to 101, the Rhind Papyrus contains a table of the quotients of the numbers 1 through 9 divided by 10. These are used to solve such problems as determining the equal division of 6 loaves of bread among 10 persons. The scribe obtains the correct answer, $\bar{2}\ \overline{10}$, from his table, and then verifies his answer by multiplying by 10:

$$\begin{array}{ll} 1 & \bar{2}\ \overline{10} \\ /\ 2 & 1\ \bar{5} \\ 4 & 2\ \bar{3}\ \overline{15}\ \text{(since } 2/5 = \bar{3}\ \overline{15}) \\ /\ 8 & 4\ \bar{3}\ \overline{10}\ \overline{30}\ \text{(since } 2/15 = \overline{10}\ \overline{30}) \\ \text{(Total)} & 6\ \text{(since } \bar{5}\ \overline{10}\ \overline{30} = \bar{3}) \end{array}$$

THE EGYPTIANS

Several other problems in the RMP are concerned with the quantity of grain needed to produce a required quantity of bread or beer. The quotient of these two quantities was called the *pesu*:

$$\text{pesu} = \frac{\text{number of loaves of bread (or jugs of beer)}}{\text{volume of grain required for production}}$$

Grain volume was measured in units called *hekats*. In one problem, 1000 loaves of pesu 10 (requiring 100 hekats of grain) are to be exchanged for a number of loaves of pesu 20 and an equal number of loaves of pesu 30. What is this number? In modern terms, the problem calls for solution of the equation

$$x/20 + x/30 = 100$$

Noting that $\overline{20}\ \overline{30} = \overline{12}$, Ahmes determines that there will be 1200 loaves of each kind.

Other problems are stated more abstractly. Many of these also involve linear equations in one unknown, such as "A quantity and its seventh part make 19. What is "the quantity?" Modern students would write

$$x + (1/7)x = 19$$

The scribe tentatively chooses the wrong answer, x = 7, and notes that 1 $\overline{7}$ times 7 equals 8. Since he wants 19 and not 8, his tentative answer must be multiplied by the number of times 8 divides 19. As you can check, the quotient of 19 by 8 is 2 $\overline{4}$ $\overline{8}$, and so the correct answer is the product of 7 by 2 $\overline{4}$ $\overline{8}$, which the scribe shows is 16 $\overline{2}$ $\overline{8}$. This technique, in which the correct answer is found as the appropriate multiple of an initial wrong guess, is usually called the "method of false position."

In other problems, the scribe used division. In one example, Ahmes seeks a quantity which when added to its $\overline{2}$ and its $\overline{4}$ becomes 10. We would write this as "Find x so that $x + x/2 + x/4 = 10$." Ahmes divides 1 $\overline{2}$ $\overline{4}$ into 10 and obtains 5 $\overline{2}$ $\overline{7}$ $\overline{14}$. He then *checks* (by calculation) that this number does have the property stipulated in the problem (see Exercise 10).

This was typical. The Egyptians often "proved" the correctness of a result by checking that their answer satisfied the conditions of the problem. This is quite different from the idea of proof as it was understood to the Greeks in Euclid's time, but we must bear in mind that the word "proof" has meant different things to mathematicians in different ages.

Nowadays, it is commonly believed that a proof must be symbolic, and that a handful of specific numerical examples is insufficient to establish a general fact. However, if the numerical values are typical, and if the method clearly generalizes to all other pertinent values, then an illustrative example may serve as a form of proof.

What amounts to the solution of a quadratic equation in one unknown is found in the Berlin Papyrus. Here it is stated that a square and a second square whose side is $\bar{2}\ \bar{4}$ times that of the first square have a combined area of 100. In applying the method of false position, the scribe tentatively chooses 1 and $\bar{2}\ \bar{4}$ as the sides of the two squares. These yield a combined area of $1^2 + (\bar{2}\ \bar{4})^2 = 1\ \bar{2}\ \overline{16}$, which is the square of $1\ \bar{4}$ (check this). He then divides $1\ \bar{4}$ into 10 to find the factor, 8, by which his false answers must be multiplied to give the correct answers of 8 and 6.

Before turning to geometry, we mention briefly that several problems in the Rhind and other papyri involve geometric and arithmetic progressions, whose properties were well known to the Egyptians. In the RMP, Ahmes poses the problem of dividing 10 hekats of barley into 10 shares in arithmetic progression with common difference $\bar{8}$. He correctly finds all ten terms, and his calculation shows that he understood the modern formula

$$\ell = S/n + (n - 1)d/2$$

in which ℓ is the last or highest term, S is the sum, n is the number of terms, and d is the common difference. It is remarkable that this could have been discovered without the benefit of any algebraic notation.

For the Egyptians, as for the Babylonians, geometry was merely applied arithmetic. Areas of triangles, rectangles, and trapezoids were correctly computed, while the area of the general quadrilateral was often reckoned as the product of the average of one pair of opposite sides times the average of the other pair—the same approximation used by the Babylonians.

The Egyptians calculated the area of a circle by squaring 8/9 of the diameter. In Problem 50 of the RMP, Ahmes assumes a circle of

THE EGYPTIANS

diameter d = 9 *khet* (about 469 meters) and finds its area as follows (Gillings [15]):

> Take away $\overline{9}$ of the diameter, namely 1.
> The remainder is 8.
> Multiply 8 by 8.
> It makes 64.
> Therefore it contains 64 *setat* (square khets) of land.
> Do it thus.
>   1   9
>   9   1
> The remainder is 8.
>   1   8
>   2   16
>   4   32
>   /8   /64
> The area is 64 setat.

By equating $(8d/9)^2$ and the correct expression $\pi d^2/4$, we see that the Egyptians approximated π as

$$\pi \approx 4(8/9)^2 = 3\ 13/81 = 3.16049\ldots$$

which is accurate to within about 0.6%.

In Problem 48 of the RMP, Ahmes has sketched what many have interpreted as a circle and a circumscribed square, but which (in view of much better drawings of circles elsewhere in the RMP) may also be interpreted as a square in which a (non-regular) octagon has been constructed by joining adjacent points of trisection of the square's sides (Fig. I-7). The convenient choice of 9 as the side of the

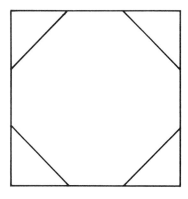

Figure I-7.

square lends support to this conjecture. By dividing his square into 9 or perhaps 81 smaller squares, and counting how many of these were covered by the octagon (which closely approximates the inscribed circle), the scribe might have found that the octagon is equivalent in area to a square of side $\sqrt{63} \approx \sqrt{64} = 8$ (cf. Gillings). This may have been the origin of the Egyptian estimate for π and would explain why this approximation was a rational square.

Several problems in the RMP deal with the volumes of rectangular and cylindrical granaries. These are correctly computed as the area of the base multiplied by the height.

A surprising accomplishment of the Egyptians was the discovery of the formula for the volume of a truncated square pyramid. This appears in Problem 14 of the Moscow Mathematical Papyrus, written about 1850 B.C.*

> Method of calculating a truncated pyramid.
> If it is said to thee, a truncated pyramid of 6 cubits in height,
> Of 4 cubits of the base, by 2 of the top,
> Reckon thou with this 4, squaring. Result 16.
> Double thou this 4. Result 8.
> Reckon thou with this 2, squaring. Result 4.
> Add together this 16, with this 8, and with this 4. Result 28.
> Calculate thou $\overline{3}$ of 6. Result 2.
> Calculate thou with 28 twice. Result 56.
> Lo! It is 56! Thou has found rightly.

A copy of the scribe's sketch, and the accompanying numbers (transliterated) is given in Figure I-8.

The solution is a particular case of the correct formula for the volume:

$$V = \frac{h}{3}(a^2 + ab + b^2)$$

where a and b are the sides of the base and top, and h is the height. In the example, a = 2, b = 4, and h = 6.

Ostensibly, the scribes must have known how to find the volume of a complete pyramid, but how they were able to deduce from this the volume of the frustum remains a mystery. It is

---

*Gilling's translation of Struve's German translation in *Quellen und Studien*, Series A, vol. 1, Berlin, 1930.

THE EGYPTIANS

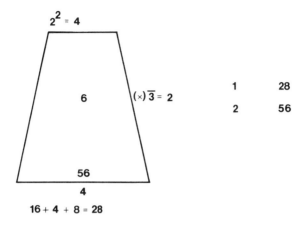

Figure I-8.

generally believed that the algebraic manipulations needed to obtain the result as the difference of a whole pyramid and a smaller pyramid cut from the top would have been beyond their capabilities (Exercise 12), although it is conceivable that the Egyptians were in some way influenced by Mesopotamian algebra. Many believe the Egyptians discovered the formula by a theoretical argument involving dissection of a truncated pyramid into several geometric objects whose volumes were easily computed. However the discovery was made, it was undoubtedly one of the supreme mathematical achievements of antiquity.

Other problems concerning pyramids dealt with what was termed the *seked,* which was a measure of the slope of the pyramid's sides. In modern terminology, the seked was the cotangent of the angle between the pyramid's base and a side. (Actually, it was seven times the cotangent because of the units of measurement employed.) Problem 56 of the RMP is typical (Gillings [15]):

> Example of reckoning a pyramid.
> Height 250, base 360 cubits.
> What is it's seked?
> Find $\bar{2}$ of 360, 180.
> Divide 180 by 250, $\bar{2}\ \bar{5}\ \overline{50}$ cubit.
> Now a cubit is 7 palms.
> Then multiply 7 by $\bar{2}\ \bar{5}\ \overline{50}$.

Ahmes then performs the indicated multiplication and obtains 5 $\overline{25}$

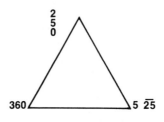

Figure I-9.

palms as the seked. His sketch appears in Figure I-9. A cubit is about 52.3 cm.

One of the greatest unsolved mysteries of Egyptian mathematics is the correct interpretation of Problem 10 of the Moscow Papyrus. This problem deals with the surface area of a basket, which probably had one of the following two shapes:

a. a hemisphere of diameter $4\,\bar{2}$;
b. a semicylinder, formed by halving a circular cylinder of diameter = height = $4\,\bar{2}$ by a plane containing the axis (Fig. I-10).

The computational procedure given by the scribe is consistent with either possibility, and Egyptian baskets were commonly made of both shapes.

Figure I-10.

THE EGYPTIANS

Let $d = 4\,\overline{2}$. In the original hieratic, the scribe first computes $8/9 \times 8/9 \times 2d$ and then multiplies the result by d. Since $4(8/9)^2$ is the Egyptian value for π, the scribe's answer, 32, is consistent with the modern formula

$$A = \pi d^2 /2$$

for the area of a hemisphere. However, if the last factor d is interpreted as the height of a cylinder, then the answer is equally consistent with the formula

$$A = \pi dh/2$$

for the lateral area of a semicylinder.

Difficulties arise in the interpretation of the scribe's poor handwriting and disagreement on the meaning of certain key words. The scribe may have made errors in copying from another document. If the Egyptians really had discovered the formula for the area of a sphere—some 1600 years before Archimedes—it would have been a remarkable achievement, of which we can only guess the method of discovery. Perhaps it was a lucky guess prompted by the observation that the amount of material required to make a hemispherical basket seems to be about twice the material needed for the circular lid.

Although the Pythagorean Theorem was well known to the Babylonians, there is no clear-cut documentary evidence that the Egyptians knew of even the simplest case of this theorem, the 3 - 4 - 5 triangle. Statements to the contrary, found in many textbooks on the history of mathematics, have their origin in the unsubstantiated theory that the surveyors, or "rope-fasteners," who laid out the Egyptian temples and pyramids constructed right angles by means of a cord divided into 3 + 4 + 5 segments by equally spaced knots.

There is also little basis in fact for such legends as that the perimeter of the Great Pyramid's base is exactly 2π times its height. (I notice that the length of my desk is very nearly π times the width of the middle drawer, but I doubt any mystical intent on the part of its manufacturer.) The often repeated tales of a high level of Egyptian applied science are just not borne out by the records.

Nevertheless, there is much to admire. With little more than knowledge of the "times 2" and "times $\bar{\bar{3}}$" tables, the Egyptians carried their calculating art to an impressive level. Of course calculation per se is not mathematics in the modern sense, but the development of algebraic techniques and the discovery of geometric relationships do qualify the royal scribes of the Middle Kingdom as mathematicians, especially when they are viewed in the context of their culture and civilization.

Exercises I-2

1. Deduce the following identities from Eqs. (6) (do not use the method of lowest common denominator):

   a. $\bar{4} + \bar{6} = \bar{3} + \overline{12}$
   b. $\bar{4} + \overline{12} = \bar{3}$
   c. $\bar{3} + \bar{4} = \bar{2} + \overline{12}$
   d. $\bar{6} + \bar{8} + \overline{24} = \bar{3}$

2. (a) Express 1 as the sum of three distinct unit fractions. (b) Use (a) to show that twice any unit fraction may always be written as the sum of four or fewer distinct unit fractions.

In the following problems, passages translated from the RMP have been adapted from [8].

3. To demonstrate that $2 \div 37 = \overline{24}\ \overline{111}\ \overline{296}$, Ahmes uses the sequence $\bar{\bar{3}}\ \bar{3}, \bar{6}, \ldots$ to find a multiple of 37 which is less than 2. He then notes how much is lacking to complete 2.

(What part is 2 of) 37? (The answer:)
$\overline{24}$ (of 37 is) 1 $\bar{2}\ \overline{24}$, $\overline{111}$ (of 37 is) $\bar{3}$, $\overline{296}$ (of 37 is) $\bar{8}$.
(Computation:)

1	37	1 37	1 37
$\bar{\bar{3}}$	24 $\bar{3}$	2 74	2 74
$\bar{3}$	12 $\bar{3}$	/ 3 111	4 148
$\bar{6}$	6 $\bar{6}$	Remains now $\bar{8}$	/ 8 296 (gives) $\bar{8}$
$\overline{12}$	3 $\overline{12}$		
/ $\overline{24}$	1 $\bar{2}\ \overline{24}$		

There remains $\bar{3}\ \bar{8}$

a. Using Exercise 1, show that $(1\ \bar{2}\ \overline{24}) + (\bar{3}\ \bar{8}) = 2$.

THE EGYPTIANS

b. Using the Egyptian method illustrated above, show that $2 \div 23 = \overline{12}\ \overline{276}$. (Hint: wherever possible, make use of Eqs. (6c) and (6d), and Exercise 1.)

c. Show that $2 \div 41 = \overline{24}\ \overline{246}\ \overline{328}$.

4. In attempting to verify a unit fraction identity, such as

$$\overline{3}\ \overline{10}\ \overline{15} = \overline{2}$$

an Egyptian scribe may have reasoned this way: "Both sides of my identity are the same provided they produce the same result when applied to (i.e., multiplied by) the same number. Therefore, to establish the identity, I will apply both sides to some convenient number—perhaps the largest denominator present—and see if the results are equal."

Thus, to prove the identity above, the scribe would apply each of his unit fractions on the left to 15 and write the products—this was often done in red ink—below the corresponding fractions:

$$\overline{3}\quad \overline{10}\quad \overline{15}$$
$$5\quad 1\,\overline{2}\quad 1$$

These latter numbers are sometimes called *red auxiliaries*. Since their sum is $7\ \overline{2}$, the same as $\overline{2}$ times 15, the identity is verified.

Using this method, show that

a. $\overline{5}\ \overline{10}\ \overline{30} = \overline{3}$
b. $\overline{7}\ \overline{14}\ \overline{28} = \overline{4}$
c. $\overline{25}\ \overline{50}\ \overline{150} = \overline{15}$
d. $\overline{15}\ \overline{25}\ \overline{75}\ \overline{200} = \overline{8}$ (Apply to 200)

5. Red auxiliaries are used also in the calculation of $2 \div 35 = \overline{30}\ \overline{42}$ in the RMP:

$$\begin{array}{ccc} \overline{35} & \overline{30}\ 1\ \overline{6} & \overline{42}\ \overline{\overline{3}}\ \overline{6} \\ 6 & 7 & 5 \end{array}$$

$$\begin{array}{cc} /\ \overline{30} & 1\ \overline{6} \\ /\ \overline{42} & \overline{\overline{3}}\ \overline{6} \end{array}$$

The fractions $\overline{35}$, $\overline{30}$, and $\overline{42}$ are being applied to 210. The method apparently is this: to find $2 \div n$, choose a convenient multiple of n, kn, such that 2k can be written as a sum of distinct divisors of kn. For n = 35, k = 6 and

$$2/35 = 12/210 = 7/210 + 5/210 = \overline{30} + \overline{42}.$$

Use this method to derive

a. $2/77 = \overline{44}\ \overline{308}$
b. $2/55 = \overline{30}\ \overline{330}$
c. $2/91 = \overline{70}\ \overline{130}$

6. Problem 57 of the RMP asks for the height of a pyramid with base 140 cubits and *seked* 5 palms 1 finger (i.e., 5 $\overline{4}$ palms, since one palm equals four fingers). Solve this problem. (One cubit equals 7 palms.)

7. Problem 52 of the RMP deals with the area of a trapezoidal parcel of land:

> Example of making a cut-off triangle of land. If is said to three, A cut-off triangle of land of 20 *khet* on the side of it, 6 *khet* in the base of it, 4 *khet* on the cut-off; what is the area of it?
> Add thou the base of it to the cut-off; there becomes 10. Make thou $\overline{2}$ of 10, namely, 5, for the giving of the rectangle of it. Make thou the multiplication: 20 up to times 5; there become 10(0); the area of it this is.

(A *khet* is 100 cubits.) Express the above procedure in terms of an algebraic equation. What do you suppose is the meaning of the phrase "for the giving of the rectangle of it."? (Egyptologists do not all agree on the literal meaning of the word translated above as "side." Possibly, the correct translation should be "height" or "altitude.")

8. In RMP Problems 41 and 42, the volume of a cylindrical granary is calculated as the area of the base times the height. Since

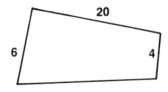

THE EGYPTIANS

the Egyptians approximated π by $4(8/9)^2$, their computation amounted to implementation of the formula

$$V = (\tfrac{8}{9}d)^2 h$$

where d and h are the diameter and height, given in cubits. Since grain volume was usually measured in units called *khar*, rather than in cubic cubits, the scribe multiplied the result from this formula by $1\ \bar{2}$, since one cubit equals $1\ \bar{2}$ *khar*.

In RMP 43 however, Ahmes uses a procedure which gives the volume in *khar* directly. After correction of certain apparent scribal errors (in accordance with a suggestion of Gillings), the translation reads:

> A granary round of diameter 8 cubits and height 6 (cubits); what is its content in grain? The doing as it occurs:
> (Add to the diameter its $\bar{3}$); becomes it $10\ \bar{\bar{3}}$. Multiply $10\ \bar{\bar{3}}$ times $10\ \bar{\bar{3}}$; becomes it $113\ \bar{3}\ \bar{9}$. Make the multiplication: $113\ \bar{3}\ \bar{9}$ up to times 4; this is $\bar{3}$ of the 6 cubits, which is the height; becomes it: $455\ \bar{9}$; the content of it, this is, in *khar*.

Write down, in terms of d and h, the formula being implemented in the passage above, and verify in general that it does give $1\ \bar{2}$ times the result of the previous formula.

9. A clever scribe would undoubtedly have mastered a number of useful tricks for reducing sums of unit fractions. Gillings suggests that the following principle, based on the identity

$$\frac{1}{n} + \frac{1}{kn} = \frac{k+1}{kn} = 1/(\frac{kn}{k+1})$$

was probably well known:

> If the denominator of one of two unit fractions is k times that of the other, then the denominator of their sum is found by dividing k + 1 into the larger denominator, provided the quotient is integral.

Use this principle to reduce each of the following sums to a single unit fraction:

(a) $\bar{4}\ \overline{12}$ (b) $\overline{15}\ \overline{60}$
(c) $\overline{14}\ \overline{21}\ \overline{42}$ (d) $\overline{21}\ \overline{42}\ \overline{84}$

10. In problem 34 of the RMP, Ahmes seeks a quantity satisfying the equation $x + \bar{2}x + \bar{4}x = 10$. He solves this problem by dividing the sum of the coefficients, $1\ \bar{2}\ \bar{4}$, into 10.

```
     / 1           1 2̄ 4̄
       2           3 2̄
     / 4           7
     / 7̄           4̄
       4̄ 28̄        2̄
     / 2̄ 14̄        1
       Total       5 2̄ 7̄ 14̄
```

(Line 5 in his division was probably suggested by the identity $2 \div 7 = \bar{4}\ \bar{28}$.)

The scribe's "proof" of his solution then begins as follows:
Example of poof:

```
     / 1       5 2̄  7̄ 14̄
     / 2̄       2 2̄ 4̄   14̄ 28̄
     / 4̄       1 4̄ 8̄   28̄ 56̄
```

Total $9\ \bar{2}\ \bar{8}$, Remains $\bar{4}\ \bar{8}$

$9\ \bar{2}\ \bar{8}$ is the total of the numbers written within the closed curve (this curve does not appear on the papyrus). Ahmes is apparently saying that to complete his "proof," it suffices to show that the sum of the remaining numbers, $\bar{7}, \bar{14}, \bar{14}, \bar{28}, \bar{28}$, and $\bar{56}$ is equal to $\bar{4}\ \bar{8}$. Show this by means of the technique of red auxiliaries described in Exercise 4.

11. In RMP 37, Ahmes seeks a quantity x satisfying the equation

$$3x + \frac{1}{3}x + \frac{1}{3}(\frac{1}{3})x + \frac{1}{9}x = 1.$$

He finds the quantity, $\bar{4}\ \bar{32}$, by dividing the sum of the coefficients, $3\ \bar{2}\ \bar{18}$, into 1:

THE EGYPTIANS

```
           1    1                  1    3 2̄ 1̄8̄
           2̄    2̄                  2̄    1 2̄ 4̄ 3̄6̄
           3̄    3̄               / 4̄    2̄ 4̄ 8̄ 7̄2̄
3̄ of 3̄    3̄    9̄                 8̄    4̄ 8̄ 1̄6̄ 1̄4̄4̄
           9̄    9̄                 1̄6̄   8̄ 1̄6̄ 3̄2̄ 2̄8̄8̄
Total          3 2̄ 1̄8̄         / 3̄2̄   1̄6̄ 3̄2̄ 6̄4̄ 5̄7̄6̄
                                  Total 1

                 2̄ 4̄ 8̄ 7̄2̄ 1̄6̄ 3̄2̄ 6̄4̄ 5̄7̄6̄
                      8 36 18  9   1
                      Total 8̄  72
```

a. Verify all steps in the calculation above and explain the use of the red auxiliaries.
b. "Prove" in the manner of Ahmes that $\bar{4}\ \bar{32}$ satisfies the original equation above. Use red auxiliaries where needed.

12. The volume of a truncated square pyramid may be realized as the difference between the volumes of two complete pyramids (see the figure on this page). Use this to derive the Egyptian formula,

$$V = \frac{h}{3}(a^2 + ab + b^2)$$

for the volume of a truncated pyramid of height h and square bases of sides a and b. (Recall: the volume of a complete pyramid equals one-third the base times the height. Use similar triangles to eliminate k.)

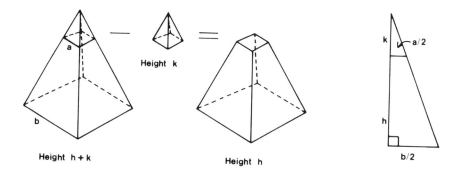

Height h + k   Height k   Height h

If you know some calculus, find the volume by evaluating the integral

$$V = \int_0^h A(y)\, dy$$

where $A(y) = [b + (y/h)(a - b)]^2$ is the area of the square cross section perpendicular to the pyramid's altitude at height y above the lower base. [Note that $A(0) = b^2$, the area of the lower base, while $A(h) = a^2$, the area of the upper base.]

II

GREEK GEOMETRY

> *He studied and nearly mastered the six books of Euclid since he was a member of Congress.*
>
> *He began a course of rigid mental discipline with the intent to improve his faculties, especially his powers of logic and language. Hence his fondness for Euclid, which he carried with him on the circuit till he could demonstrate with ease all the propositions in the six books; often studying far into the night, with a candle near his pillow, while his fellow-lawyers, half a dozen in a room, filled the air with interminable snoring.*
>
> —Abraham Lincoln
> (*Short Autobiography*)

INTRODUCTION

Our knowledge of ancient Greek mathematics (6th century B.C. through the 6th century A.D.) is based on a considerably broader foundation than our knowledge of the Babylonians and Egyptians. Extant are most of the books of a large historical and mathematical compendium called the *Collection* (c. 320 A.D.), by Pappus of Alexandria, and a *Commentary* (on Book I of Euclid's *Elements*) by the philosopher Proclus (410-485 A.D.). This *Commentary* includes a summary of a lost history of mathematics by one Eudemus of Rhodes (c. 320 B.C.), a student of Aristotle. We also have fragments of the original writings of a number of authors and various critical works written some centuries later. There are also various works based on Arabic translations of the Greek originals.

Nevertheless there is much we do not know about the lives of many of the men we shall discuss, and many of the legends about their characters or views are probably apocryphal.

In contrast to the Babylonians and Egyptians, the Greeks employed logical demonstration rather than trial-and-error experimentation: deduction instead of induction. The Greeks also evolved the first complete scientific theories of the heavens, and we shall discuss the geometric content of these in Section 7.

1. THALES OF MILETUS

Born in the Ionian Greek colony on the western coast of what is now called Asia Minor, Thales (c. 632-546 B.C.) was a well-traveled businessman, merchant, and public figure. He frequently traveled in Egypt, and in his leisure, studied the geometry and astronomy which he picked up in discussions with the priests and other learned men of his time.

Judging from the many legends surrounding his name, he was well known for his shrewdness and wit. For example, it is said that in anticipation of a large olive crop, he bought up all of the olive presses to be had and cornered the market in olive oil.

When only middle-aged he retired from public life and devoted himself to the study of philosophy and science. He founded the Ionian school, the earliest Greek school of mathematics and philosophy. There is little doubt that Thales was influenced by Babylonian mathematics and astronomy.

According to some accounts, Thales was the first to see the need for logical demonstration rather than trial-and-error experimentation. Though his propositions were not arranged in a logical sequence, nor were they very deep, the proofs were nevertheless deductive, proceeding from a few "self-evident" suppositions to necessarily following conclusions.

Some of the propositions ascribed to Thales are: the base angles of an isosceles triangle are congruent; when two lines intersect, the vertical angles are congruent; two angles and the included side determine a triangle; triangles with corresponding angles congruent have proportional sides; and an angle inscribed in a semicircle is right (considered one of his most remarkable results). Legend has it that

THALES OF MILETUS

Thales sacrificed an ox to the gods upon discovering or demonstrating this last proposition.

Elementary as these propositions were, they were here stated in general abstract terms and proven logically for the first time. In place of specific material disks, squares, and triangles, Thales was considering ideal geometric shapes.

He taught that a year contains 365 days, and is said to have predicted the solar eclipse of 585 B.C., probably by making use of the fact that solar eclipses recur at regular intervals. While in Egypt he is said to have found the height of the Great Pyramid by comparing its shadow with that of a vertical stick, a use of similar triangles apparently unknown to the Egyptians (see Exercise 1). For such feats as these he was known by succeeding generations as one of the Seven Sages of Greece.

Although he was more of an astronomer than a geometer, and although the school which he founded turned more to questions of philosophy, Thales is generally considered largely responsible for converting mathematics from an inductive to a deductive discipline.

Exercise II-1

Hieronymus, a pupil of Aristotle, says that Thales determined the height of a pyramid by measuring the shadow it cast at the moment a man's shadow was equal in length to his height. A later version of this, given by Plutarch, says that Thales set up a stick and then made use of similar triangles. Let ℓ be the (known) height of a vertical stick, and let h be the (unknown) height of the pyramid (see the diagram on this page). Then by similar triangles, $h/S = \ell/s$ and so

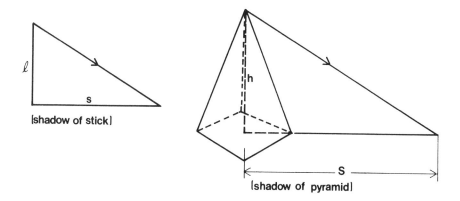

h = ℓS/s. Unfortunately, we are unable to measure S directly, because we cannot get inside the solid pyramid to locate the center of its square base. Both versions of the above legend fail to mention this difficulty.

Devise a method of your own, based on similar triangles and independent of latitude and time of year, for determining the pyramid's height from *two* shadow observations made at different times.

2. THE PYTHAGOREAN SCHOOL

Pythagoras (c. 570-500 B.C.) studied first under a disciple of Thales, and later in Egypt. He traveled and lectured in Asia Minor, migrated to Sicily, and eventually settled in Crotono in southern Italy. There he opened up popular schools of mathematics and philosophy. His lectures were heavily attended. Even women, forbidden by law from attending public meetings, flocked to hear him. One of these was the beautiful daughter of his patron, Milo. Although much younger, she married Pythagoras and later wrote his biography, unfortunately lost.

Pythagoras divided his students into two classes: probationary members and full-fledged Pythagoreans. The latter were formed into a secret mathematical, philosophical, and political society, whose members shared all things in common, lived under strict discipline and veiled their knowledge from the outside world. (Pythagoras, himself, did not publish.) Hippasus (c. 470 B.C.), a later disciple, is said to have been drowned for violating his oath by publicly boasting of his discovery of the dodecahedron as one of the five regular solids, although a more political motive may have been responsible.

The secrecy and political influence of the Pythagoreans brought on the hatred of various factions, and (according to varying accounts) at the instigation of his opponents, Pythagoras and/or many of his followers were murdered by a mob.

Primarily a philosopher and moral reformer, his concept of a liberal education centered around a "quadrivium" (literally "four roads") of four basic areas: numbers absolute (arithmetic), numbers applied (music), magnitudes at rest (geometry), and magnitudes in motion (astronomy).*

---

*The classical "seven liberal arts" consisted of the quadrivium auugmented by a "trivium" made up of grammar, logic, and rhetoric.

THE PYTHAGOREAN SCHOOL

In geometry, Pythagoras may have been the first to arrange large sets of propositions in a logical order. Eudemus (according to Proclus) wrote,

> ". . . he changed the study of geometry into the form of a liberal education, for he examined its principles to the bottom and investigated its theorems in an intellectual manner."

It is difficult to separate the discoveries of Pythagoras from those of his followers, but it is believed he knew most of the results that can be found in Euclid's *Elements* (Section 4) on triangles, parallels, and parallelograms, including the theorem that the sum of the angles in a triangle equals two right angles. Most likely, he proved the theorem that bears his name.

The Pythagoreans knew that there were five regular solids inscribable in a sphere, and that a plane could be "tiled" with squares, equilateral triangles, or regular hexagons. There is reason to believe that the later Pythagoreans knew that the side and diagonal of a square were incommensurable (i.e., their ratio cannot be expressed as a quotient of whole numbers: $\sqrt{2}$ is irrational).

The Pythagoreans elevated arithmetic above the needs of the merchant to the level of a philosophy. They believed that an understanding of order in the universe was to be derived from the science of numbers. For instance, the lengths of vibrating strings giving a note, its fifth, and its octave are in the ratio $4:3:2$, and the distances of the planets from the earth were believed to form an integer progression. Some believed the planets and other celestial bodies gave out musical tones (inaudible to mortal man), hence the expression "the music of the spheres." Another version of the Hippasus legend states that Hippasus was expelled from the Pythagorean society for revealing the discovery of irrational numbers, a discovery devastating to the Pythagorean philosophy, which sought to explain everything in terms of whole numbers and their ratios.

Gradually, the society became less secretive as its members scattered and began to produce written works. By the fourth century B.C., the center of intellectual activity had shifted to Athens.

One of the most notable members of the Pythagorean school was Archytas (c. 400 B.C.), who treated mechanics with the aid of geometry. He is said to have worked out the theory of the pulley and constructed ingenious mechanical devices and toys, including a mechanical bird. He taught that the earth was a sphere rotating around its axis in 24 hours, and that the heavenly bodies moved around it. He also gave a construction for finding the side of a cube double the volume of a given cube (see Exercises 3 and 4). The significance of this problem will be discussed further in the next section and the exercises. His students founded the Athenian school of mathematics.

Exercises II-2

1. The Hindu mathematician Bhaskara (c. 1150 A.D.) used the diagram on this page to "prove" the Pythagorean Theorem. In this diagram, the square on the hypotenuse is cut into four triangles, each congruent to the given triangle, plus a square with side equal to the difference of the legs of the triangle. Bhaskara simply drew the figure and offered no further explanation. Supply a proof. (Hint: Show that the figure can be constructed starting from the inner square. Show $c^2 = a^2 + b^2$.)

2. Let $\triangle ABC$ be a right triangle with right angle at C. The Pythagorean Theorem (the assertion that $AB^2 = AC^2 + BC^2$) was viewed by the Greeks as a statement about areas of squares (Figure A). Now the areas of any two similar figures are in proportion to the squares of corresponding linear dimensions. For example, the areas of two semicircles are proportional to the squares of their diameters. Consequently, an assertion equivalent to the Pythagorean Theorem is obtained when, in the quoted sentence at the bottom of page 51, the word "square" is replaced by "semicircle" (Figure B), or by "equilateral triangle," or by "regular pentagon." In fact, any type of plane figure can be used, as long as the figures erected on all

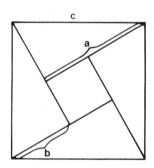

THE PYTHAGOREAN SCHOOL

three sides of △ ABC are similar and the triangle's sides are corresponding parts of these figures.

Indeed we can even use right triangles similar to △ ABC itself! (In this case, the figure erected on $\overline{AB}$ is △ ABC itself.) Give a (nearly trivial) proof of the Pythagorean Theorem based on these ideas (see Figure C).

3. Given two positive numbers a and b, positive numbers x and y are said to be *two geometric means* between a and b if

$$\frac{a}{x} = \frac{x}{y} = \frac{y}{b}$$

Show that if $a = 2b$, then $y^3 = 2b^3$. (Since a cube of side y has twice the volume of a cube of side b, the existence of a geometric construction method for the two geometric means between any two given lengths would solve the problem of constructing a cube double the volume of a given cube. The following exercise describes such a construction, due to Archytas.)

4. Let a and b, with $b < a$, be the lengths of two given segments. Here is an analytic version of Archytas's construction of the two geometric means between a and b (see Exercise 3). We use the following facts from solid analytic geometry: if a curve $g(x,z) = 0$ in the xz-plane is revolved about the z-axis, then the resulting surface of revolution has equation $g([x^2 + y^2]^{1/2}, z) = 0$; when the curve is revolved about the x-axis, the generated surface has equation $g(x, [y^2 + z^2]^{1/2}) = 0$.

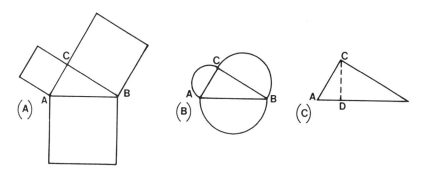

"The area of the square erected on the hypotenuse equals the sum of the areas of the squares erected on the two legs."

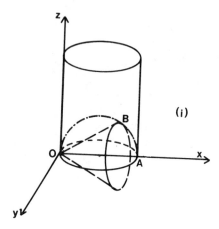

(i)

First, construct a right circular cylinder of diameter a, whose base is the circle in the xy-plane with center (a/2,0,0), as in Figure (i). This cylinder has equation $(x - a/2)^2 + y^2 = a^2/4$, or

$$x^2 + y^2 = ax$$

Upon $\overline{OA}$ as diameter, construct a semicircle in the xz-plane. This semicircle has equation $x^2 + z^2 = ax$, and, when revolved about the z-axis, generates a second surface, with equation

$$x^2 + y^2 + z^2 = a(x^2 + y^2)^{1/2}$$

[not shown in Figure (i)].

Finally, let $\overline{OB}$ be a chord of length b in the semicircle above. By similar triangles [Fig. (ii)], point $C = (x,0,z)$ on this chord satisfies $a/b = (x^2 + z^2)^{1/2}/x$. When OB is revolved about the x-axis, we obtain the cone with equation

$$\frac{a}{b} = \frac{(x^2 + y^2 + z^2)^{1/2}}{x}$$

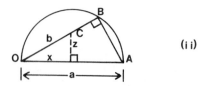

(ii)

THE PYTHAGOREAN SCHOOL

Let P = (x,y,z) be a point common to all three surfaces just described. Let $\rho = OP = (x^2 + y^2 + z^2)^{1/2}$ and let $r = (x^2 + y^2)^{1/2}$ = the distance from O to the projection of P onto the xy-plane. Show that

$$\frac{a}{\rho} = \frac{\rho}{r} = \frac{r}{b}$$

3. THE ATHENIAN SCHOOL

Following the decline of the Pythagorean school, the center of Greek learning shifted to Athens. It was here that the famous Academy of Plato attracted many of the greatest philosophical and mathematical minds of the fourth century B.C. A companion of Archytas, as well as a pupil of Socrates, Plato stressed the interplay of mathematics and philosophy. Like the Pythagoreans, he believed that the secrets of the universe lay in form and number. It is said that the gates of his Academy bore the inscription "Let no one ignorant of geometry enter herein."

Because the circle and line are the most elemental, perfect, and symmetric of geometric forms, Plato objected to the use of any instruments other than compass and straight-edge in geometric construction. The 2nd century A.D. historian Plutarch indicates Plato's reaction to the use of other mechanical devices in geometry [11]:

> But what with Plato's indignation at it, and his invectives against it as the mere corruption and annihilation of the one good of geometry—which was thus shamefully turning its back upon the unembodied objects of pure intelligence to recur to sensation, and to ask help (not to be obtained without base subservience and depravation) from matter; so it was that mechanics came to be separated from geometry, and, repudiated and neglected by philosophers, took its place as a military art.

Plato stressed the importance of precise definitions and axioms, and lectured his followers and associates on the proper methods to be used in attacking mathematical problems. Although not a great mathematician himself, he was a "maker of mathematicians" and had an enormous influence on his contemporaries and successors.

Much of the work of the Athenian geometers arose out of efforts to solve what have become known as the "three classical problems" of antiquity: quadrature, or squaring, of the circle, i.e., find a construction that produces a square equal in area to a given circle; trisection of the angle, i.e., find a construction which will divide any given angle into three congruent parts; and duplication of the cube, i.e., give a construction that produces a cube of volume exactly double that of a given cube.

According to one legend, the Athenians in 429 B.C. appealed to the oracle of Apollo at Delos to stop a plague that was devastating the city. The oracle told them to double (in volume) the size of the god's cubical altar. Faithfully, the Athenians constructed a new cubical altar, with each side exactly double that of the original. Since this increased the volume eightfold, rather than doubled it, the plague continued. The duplication problem has since come to be known as the "Delian problem."

All three problems are insoluble if one is restricted to the use of straightedge and compass alone, although the proof of the impossibility did not come until modern times. (See Sawyer [40] for the angle trisection problem.)

The greatest luminary of the Athenian school was Eudoxus (c. 408-355 B.C.), who is credited with two monumental achievements in mathematics (his theory of proportions and the method of exhaustion) as well as the first astronomical theory of the heavens (Section 7).

The theory of proportion addressed the problem of incommensurables, which had become more and more perplexing since their discovery by the Pythagoreans. These quantities, which could not be expressed as ratios of integers, seemed to defy understanding. How could they be compared? Were they numbers or not?

To circumvent these difficulties, Eudoxus used the notion of *magnitude* (e.g., length, area, volume, weight, time), rather than number. Two magnitudes of the same type had some entity in common called a *ratio*. The theory of proportions appearing in Book V of Euclid's *Elements* (Section 4) is essentially that of Eudoxus. Here are Euclid's definitions of ratio and magnitude*:

> A ratio is a sort of relation in respect to size between two magnitudes of the same kind.
>
> Magnitudes are said to have a ratio to one another which are capable, when multiplied, of exceeding one another.

Thus we may compare (or form the ratio of) two magnitudes of the same kind such as two areas or two volumes, but not two magnitudes

---

*Definitions and propositions from Euclid's *Elements* are reprinted from [18].

THE ATHENIAN SCHOOL

of different kinds, such as a volume and an area. (Compare this with the algebra of the Babylonians, who had no compunction about adding or multiplying different kinds of things, such as volumes and lengths, or areas and wages.) The second definition above embodies what is known as the Axion of Archimedes, although this principle is due to Eudoxus:

> Given two (nonzero) magnitudes, a sufficiently large multiple of the smaller exceeds the larger.

Using modern symbolism, let us use a:b to denote the ratio of two magnitudes of the same kind. According to Euclid, a:b exists if and only if there exist integers m, n such that ma > b and nb > a. Equality and inequality of ratios are then defined as follows.

Definition II-1. a:b = c:d if for integers m and n,

  i. whenever ma < nb, then mc < nd,
  ii. whenever ma = nb, then mc = nd,
  iii. whenever ma > nb, then mc > nd.

Definition II-2. If there exist integers m and n such that ma > nb but mc ≤ nd, then a:b > c:d.

It is then possible to deduce the transitivity property for ratios (a:b > c:d and c:d > e:f imply a:b > e:f), and all the standard rules for manipulating proportions. For example if a:b = c:d, then a:c = b:d. Some of the proofs are requested in the exercises.

Many scholars see in the ratios of Eudoxus the seeds of Dedekind's development of the real number system in the nineteenth century. Dedekind essentially defined an irrational number as a partition of the set of all rational numbers into two disjoint non-empty classes A, B such that (1) every rational in A is less than every rational in B, and (2) there is no largest member of A and no smallest member of B. If a and b are incommensurable, then we can associate with a:b just such a partition, namely:

$$A = \left\{ \text{rationals } \frac{m}{n} \mid m\,b < n\,a \right\}$$
$$B = \left\{ \text{rationals } \frac{m}{n} \mid m\,b > n\,a \right\}$$

(see Exercise 4).

The theory of proportions forced a sharp separation between geometry and arithmetic and led to a deemphasis of number as the primal concept. Geometry became the main arena for rigorous mathematical pursuit and remained so for nearly two thousand years. As we shall see in Section 4, the Greeks of the Alexandrian period even stated algebraic equations in terms of geometric relationships. The problem of incommensurables also contributed to the appreciation of the need for a precise axiomatic foundation and careful deductive reasoning.

Eudoxus also developed a type of limit process known as the *Method of Exhaustion*. This was based on the following principle (Euclid, Book X, Prop. I):

> Two unequal magnitudes being set out, if from the greater there be substracted a magnitude greater than its half, and from that which is left a magnitude greater than its half, and if this process be repeated continually, there will be left some magnitude which will be less than the lesser magnitude set out.

An example of the principle's use is this proof (of Archimedes) that the difference between the area of a circle and that of an inscribed regular polygon can be made as small as desired by increasing the number of the polygon's sides.

In Figure II-1(a), ABCDE is a typical polygon inscribed in a circle. The shaded region (made up of circular segments) represents the difference in area between the circle and the polygon. Suppose

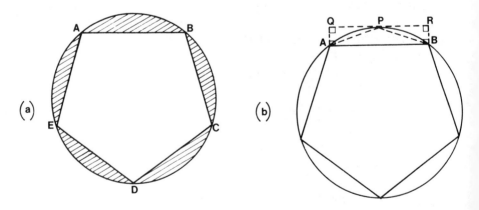

Figure II-1.

THE ATHENIAN SCHOOL

we double the number of the polygon's sides by adding as new vertices the midpoints of arcs $\widehat{AB}$, $\widehat{BC}$, $\widehat{CD}$, etc. How does this increase the polygon's area? In Figure II-1(b), we see that the addition of vertex P adjoins △APB to the previous polygon. Since

area △APB < area segment APB < area rectangle QRBA = 2 area △APB

we have

area △APB > $\frac{1}{2}$ area segment APB

A similar argument holds for the other sides of the polygon. Consequently, doubling the number of sides adds on enough area to remove more than half the difference between the area of the circle and that of the polygon. Therefore, by the exhaustion principle, repeated doubling will eventually make the difference as small as desired.

As we know, this idea of obtaining the area of a curvilinear figure as the limit of rectilinear figures (polygons or rectangles) reached fruition in the integral calculus of Newton and Leibniz in the 17th century.

One other mathematician we must note before closing this section is Menaechmus (c. 375-325 B.C.), for he was the discover of the conic sections, which he obtained not from different kinds of sections of one fixed cone, but rather by slicing a right circular cone by a plane perpendicular to an element and then varying the cone's vertex angle. If the vertex angle is acute, the section is an ellipse; for a right angle, a parabola is obtained; while for an obtuse angle, the section is (one branch of) a hyperbola (Fig. II-2). Menaechmus studied the properties of these curves and successfully used them to solve the Delian problem (Exercise 6).

Exercises II-3
1. Show (using Definition II-1) that

   a. if $a:b = c:d$ then $(a + c):(b + d) = a:b$
   b. $ma:mb = a:b$ for every positive integer m

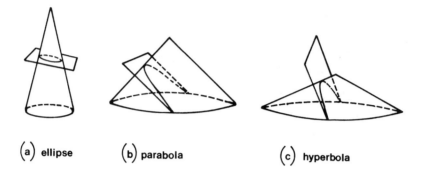

(a) ellipse (b) parabola (c) hyperbola

Figure II - 2.

2. Assume the following theorem (Euclid V, 14): if a:b = c:d, then

   a > c implies b > d
   a = c implies b = d
   a < c implies b < d

Using this and Exercise 1b, prove that

   a:b = c:d implies a:c = b:d.

3. Prove that

   a. a:b = c:d, c:d = e:f imply a:b = e:f
   b. a:b > c:d, c:d > e:f imply a:b > e:f

4. We have seen that if a and b are incommensurable, we may construct a partition of the set Q of all rational numbers into two disjoint classes A and B, defined by

$$A = \left\{ \frac{m}{n} \in Q \mid mb < na \right\}$$
$$B = \left\{ \frac{m}{n} \in Q \mid mb > na \right\}$$

Suppose that a':b' = a:b in the sense of Definition II-1, and let

THE ATHENIAN SCHOOL

$$A' = \left\{ \frac{m}{n} \epsilon\, Q \mid mb' < na' \right\}$$
$$B' = \left\{ \frac{m}{n} \epsilon\, Q \mid mb' > na' \right\}$$

Show that A = A', B = B', so that the partition (Dedekind's concept of an irrational number) depends only on the ratio of the incommensurables (Euclid's concept of an irrational number), and not on the choice of a and b giving that ratio.

5. Use the method of exhaustion to show that the difference between the area of a circle and that of a circumscribed regular polygon can be made as small as desired by increasing the number of the polygon's sides. (Hint: in the figure, A and B are midpoints of adjacent sides of a circumscribed polygon. If the number of sides is doubled by drawing tangents at the midpoints of arcs such as AB, then the excess area AMBN, shaded, is reduced by area ($\triangle$CND) = 2 area ($\triangle$CMN). Show area ($\triangle$ACM) < area ($\triangle$CNM).)

6. Show that the parabolas $x^2 = 2ay$ and $y^2 = ax$ intersect in a point (x,y) satisfying

$$\frac{2a}{x} = \frac{x}{y} = \frac{y}{a}$$

What does this have to do with the Delian problem?

4. EUCLID

After the death of Alexander in 323 B.C., his empire was disbanded. Egypt was put under the control of Ptolemy, who chose Alexandria

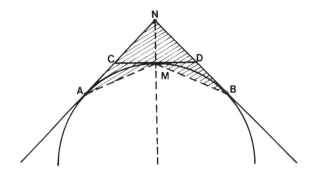

as his capital. Determined to make the city a center of learning, he constructed a university, known as the Museum (c. 300 B.C.), and recruited from Athens some of the most learned men of the time, including Euclid who became head of the mathematics department. This university became the center of Greek learning for nearly a thousand years. Its first century produced three of the greatest of Greek mathematicians, Euclid, Archimedes, and Apollonius.

Of Euclid, not much is known of his life. He became a successful teacher and settled in Alexandria. According to tradition, he was a kindly, modest man, who, when asked by Ptolemy for an easy introduction to geometry, is said to have replied "There is no royal road to geometry." (A similar comment, made to Alexander, has also been attributed to Menaechmus.)

Euclid wrote the most influential textbook in the history of civilization, the *Elements*. Divided into thirteen volumes, the *Elements* was essentially a single chain of some 465 propositions, encompassing most of the geometry, number theory, and geometric algebra of the Greeks. Although much of the text is based on the works of previous writers, many of the results and the form of the work are Euclid's. For two thousand years Euclid's *Elements* was the standard textbook in geometry. There have been at least a thousand editions, and no other work, except the Bible, has been more widely used and studied.

Euclid published several other books as well. Among these were a collection of problems known as the *Data*, which was intended to create problem-solving ability and develop originality in the student of geometry. His treatise *On Divisions* dealt with the problem of dividing a given region into parts whose areas are in a desired ratio. His *Phaenomena* was a text on spherical geometry for astronomers, and the *Optics* treated the geometry of perspective (i.e., the geometry of visual perception). There were several lost works also, including a book on geometric fallacies (*Pseudaria*), a collection of treatises on conic sections, a treatise on porisms, and a book on music. (A *porism* is a proposition giving conditions that render a particular problem either indeterminate or capable of infinitely many solutions, as for example the famous porism of Steiner (cf. Kay [23] or Ogilvy [34]). Euclid's porisms may or may not have fit this definition closely.)

EUCLID

Since so much of the next few chapters has a bearing on Euclid's work, we will briefly describe the contents of the thirteen books of his *Elements*.

Book I contains the preliminary definitions, axioms, and postulates of plane geometry. The proved propositions include the familiar results on triangles (e.g., base angles of an isosceles triangle are congruent), as well as the four congruence theorems for triangles: side-angle-side (SAS), side-side-side (SSS), angle-side-angle (ASA), and side-angle-angle (SAA). Book I contains also the theory of parallels and the Pythagorean theorem (see Appendix A).

Book II is devoted to Greek geometric algebra (the solution of algebraic problems by geometry). The Alexandrian Greeks expressed algebraic equations as relationships among lengths, areas, and volumes, rather than among numbers. For example, the Pythagorean Theorem was not seen as a statement about the squares of the numeric measures of a right triangle's sides, but as a statement about areas of squares erected on these sides (see Euclid I, 47 in Appendix A).

As an illustration of geometric algebra, consider Proposition 11 of Book II:

Euclid II, 11. (It is possible) to cut a given straight line so that the rectangle contained by the whole and one of the segments is equal to the square on the remaining segment.

If a denotes the length of the given line (segment) and if a - x and x are the lengths of the two segments into which it is to be cut or divided, then the statement may be seen as requesting a solution of the equation

$$a(a - x) = x^2$$

Euclid's solution runs as follows (see Fig. II-3, in which $\overline{AB}$ is the given segment of length a).

Construct square ABDC, and bisect $\overline{AC}$ at E. Draw $\overline{BE}$. Extend $\overline{CA}$ through A to F so that EF = BE. Construct square FAHQ.

Then Euclid goes on to demonstrate that the rectangle contained by $\overline{AB}$ and $\overline{BH}$ (i.e., a rectangle with sides congruent to these) is equal in area to the square on $\overline{AH}$. Equivalently, area (rect. HBDK)

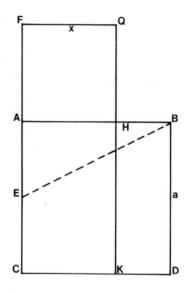

Figure II - 3.

= area (sq. FQAH). Thus H produces the desired division of $\overline{AB}$ (AH = x). Other examples of geometric algebra will appear in the exercises.

Book III is devoted to circles and chords, tangents, and angles inscribed in circles. For example,

> *Euclid III, 35.* If in a circle two straight lines cut one another, the rectangle contained by the segments of the one is equal to the rectangle contained by the segments of the other.

In brief, in the circle of Figure II-4(a), AX·XB = CX·XD. Another important result is the so-called Secant-Tangent Theorem, Euclid III, 36: in Figure II-4(b), if $\overleftrightarrow{PA}$ is tangent to the circle and $\overleftrightarrow{PBC}$ is any secant line, then $(PA)^2$ = PB·PC.

Book IV gives the Pythagorean constructions, with straightedge and compass, of the regular polygons of 3, 4, 5, 6, and 15 sides.

Book V contains the theory of proportion originated by Eudoxus (Section 3).

Book VI applies the theory of proportions to similar triangles, geometric proportions, and the solution of certain quadratic equations (e.g., x^2 = ab or a:x = x:b).

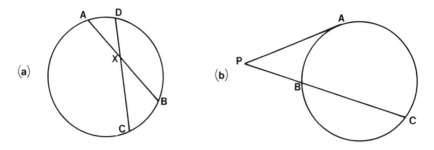

Figure II - 4.

Books VII, VIII, and IX deal largely with the theory of numbers (treated geometrically, as lengths), and will not be discussed further here.

Book X deals with incommensurables and is based on the work of Eudoxus.

Books XI, XII, and XIII contain solid geometry (lines and planes in space and volumes) and construction of the five regular polyhedra (the tetrahedron, cube, octahedron, dodecahedron, and icosahedron).

We will discuss the *Elements* further in Chapter III.

Exercises II-4

1. (a) Let r and s denote the roots of the equation $x^2 - px + q^2 = 0$, where p and q are positive. Show that $r + s = p$, $rs = q^2$, and that r and s are positive real numbers if and only if $q \leq p/2$. (b) To solve the above quadratic equation geometrically, we must find r and s from given segments of lengths p and q. That is, we must construct a rectangle equivalent (equal in area) to a given square (q^2) and having the sum of its base and altitude equal to a given line segment ($r + s = p$). Devise a suitable construction based on the figure, and show *geometrically* that for real roots to exist, we must have $q \leq p/2$.

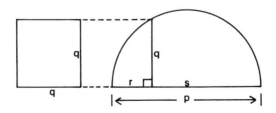

2. Can you solve (geometrically) $x^2 - px - q^2 = 0$, where p and q are positive. (Hint: use the figure below and Euclid III, 36. Denote the roots by r and -s.)

3. The *mean proportional* (or *geometric mean*) between two positive numbers a and b is the positive number x such that $a/x = x/b$.

   a. Given segments of length a and b, draw a circle of radius a + b. On any diameter $\overline{AB}$ of this circle, let C be the point such that AC = a, CB = b. Erect $\overline{CD}$ perpendicular to $\overline{AB}$, where D lies on the circle. Show that CD is the mean proportional between a and b.
   b. Show that a special case of the above gives a construction for the square root of a given length.

4. Indicate how each of the following identities might be established geometrically (i.e., by adding and/or subtracting areas in diagrams), assuming a and b are given lengths.

   (i) $(a + b)^2 = a^2 + 2ab + b^2$
   (ii) $(a - b)^2 = a^2 - 2ab + b^2$ (a > b)
   (iii) $a^2 - b^2 = (a + b)(a - b)$ (a > b)

5. It is a theorem of Euclidean geometry (essentially the content of Euclid VI,2) that if side $\overline{AB}$ of $\triangle ABC$ is partitioned into n congruent segments by the addition of n - 1 equally spaced points, then parallels to $\overleftrightarrow{BC}$ drawn through these points will partition $\overline{AC}$ into n congruent segments. Describe a geometric construction for dividing a given line segment into n congruent parts, based on this theorem.

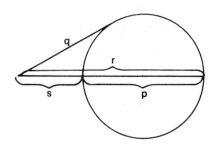

5. ARCHIMEDES

The greatest mathematician of antiquity, often ranked with Gauss and Newton among the giants of all time, was Archimedes (287-212 B.C.). A native of Syracuse on the island of Sicily, he studied for a time at the Museum in Alexandria, then returned to spend the rest of his life in Syracuse. His mathematical and scientific accomplishments were too numerous for us to consider more than a few of them.

Archimedes applied the method of exhaustion created by Eudoxus to the problem of computing areas and volumes. Here for example, is his proof that the area of a circle is one-half the product of its radius and circumference.

Let A and C be, respectively, the area and circumference of a circle of radius r. We wish to prove that $A = \frac{1}{2}rC$. Suppose this were not true. Suppose, for example, that $A > \frac{1}{2}rC$. As indicated in Section 3, the method of exhaustion can be used to produce an inscribed regular polygon whose area differs arbitrarily little from A, in particular, so little as to lie between A and $\frac{1}{2}rC$. Let the perimeter and area of such a polygon be denoted C_p and A_p. Then

$$\frac{1}{2}rC < A_p < A$$

Let r' be the apothem of this polygon, i.e., the perpendicular distance from the center to any side.

In Figure II-5, we see that A_p is the sum of the areas of n triangles such as $\triangle AOB$, each with area $\frac{1}{2}r's$ where s is the length of the polygon's side. Summing these n triangular areas, we have

$$A_p = \frac{1}{2}r'ns = \frac{1}{2}r'C_p$$

and so

$$\frac{1}{2}rC < \frac{1}{2}r'C_p$$

This is impossible because $r > r'$ and $C > C_p$.

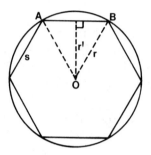

Figure II - 5.

Therefore, the initial supposition $A > \frac{1}{2}rC$ cannot hold. By a similar argument, using circumscribed polygons, Archimedes showed that $A < \frac{1}{2}rC$ is likewise impossible. Therefore $A = \frac{1}{2}rC$.

By considering both circumscribed and inscribed polygons, Archimedes was able to arrive at an excellent approximation to π. He developed an iterative procedure that would give him the perimeters of the inscribed and circumscribed polygons of 2n sides from the perimeters of those of n sides. Beginning with n = 6 and repeatedly doubling until 96-sided polygons were obtained, he was able to show that

$$3\frac{10}{71} < \pi < 3\frac{10}{70}$$

a better estimate than those obtained by either the Babylonians or the Egyptians.

By a similar use of inscribed polygons, Archimedes found the area of a parabolic segment, such as ABV in Figure II-6(a). Let M be the midpoint of chord $\overline{AB}$ and let $\overleftrightarrow{MC}$ be drawn parallel to the axis $\overleftrightarrow{VW}$ of the parabola [Fig. II-6(b)]. Archimedes knew that the tangent line to the parabola at C is parallel to $\overleftrightarrow{AB}$ and that $\overleftrightarrow{MC}$ bisects every chord parallel to $\overleftrightarrow{AB}$. $\overleftrightarrow{MC}$ is called the *diameter* of the parabolic segment. Now draw the diameters $\overleftrightarrow{M_1 C_1}$ and $\overleftrightarrow{M_2 C_2}$ of segments ACC_1 and BCC_2, respectively. Archimedes proved that

$$\text{Area } \triangle ACC_1 = \text{Area } \triangle BCC_2 = \frac{1}{8} \text{ Area } \triangle ABC$$

ARCHIMEDES

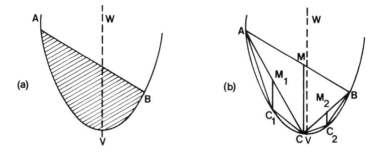

Figure II - 6.

Let δ = Area $\triangle$ ABC. We have just seen that adding the triangles $\triangle ACC_1$ and $\triangle BCC_2$ to $\triangle ABC$ gives a polygonal figure ABC_2CC_1 of area $\delta + \delta/4$ inscribed in the original parabolic segment. We may repeat this process, constructing four more triangles lying on $\overline{AC_1}$, $\overline{C_1C}$, $\overline{CC_2}$, and $\overline{C_2B}$, and giving additional area $\delta/16$. After n repetitions of this process, we have an inscribed polygon of total area

$$\delta + \delta/4 + \delta/16 \cdots + \delta/4^n \tag{7}$$

Moreover, Archimedes used the method of exhaustion to show that this polygonal area can be made arbitrarily close to the segment's area by reiterating sufficiently many times. Archimedes then gave what amounts to a proof that the sum in (7) approaches $(4/3)\delta$ as n increases indefinitely.

Although Archimedes' proofs were masterpieces of logic and clear exposition, they gave no clue as to how they were discovered in the first place. Luckily, in 1906, a 10th century manuscript was discovered containing transcriptions of several known works of Archimedes plus a hitherto unknown work called *The Method*. It was revealed in this work that Archimedes often used a mechanical method that involved mentally weighing slices of one figure against slices of another, simpler figure whose area or volume was known. The method was based on the law of the lever, which Archimedes had derived in another work: two objects suspended from opposite arms of a balance beam will be in equilibrium if their weights are inversely proportional to their distances from the fulcrum. For example, a 160 pound adult sitting four feet from the pivot point of a seesaw will exactly balance an 80 pound child sitting eight feet from the fulcrum on the other side. Both weights exert a *moment* or turning force of 640 (= 160 × 4 = 80 × 8) footpounds.

The method is best illustrated through the following example, in which we show how Archimedes obtained the formula for the volume of a sphere. Although we shall use modern symbolism, the argument is essentially the one Archimedes gave in *The Method*.

Consider a circle of radius r centered at (r,0) in the xy-plane. The equation of this circle is

$$(x - r)^2 + y^2 = r^2$$

or

$$y^2 = 2rx - x^2 \tag{8}$$

We wish to find the volume of the sphere obtained by revolving this circle about the x-axis.

To do this, we will make use of two additional solids: the cone obtained by revolving the segment $y = x$, $0 \leq x \leq 2r$, about the x-axis, and the cylinder obtained by revolving about the x-axis the circle's circumscribed square with sides parallel to the coordinate axes (see Fig. II-7).

The cone has base radius 2r and height 2r and so its volume is

$$V_{cone} = (8/3)\pi r^3$$

The cylinder has radius r and height 2r, and so its volume is

$$V_{cyl} = 2\pi r^3$$

Archimedes had derived both of these formulas earlier.

Now imagine that we slice each of these three solids by the plane perpendicular to their common axis at some x, $0 \leq x \leq 2r$. The slice of the sphere is a disk of area

$$\pi y^2 = \pi(2rx - x^2)$$

from (8). The slice of the cone is a disk of area πx^2, and the slice of the cylinder is a disk of area πr^2.

ARCHIMEDES

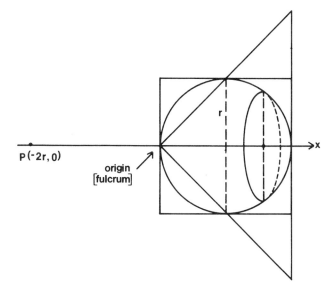

Figure II - 7.

We now pretend that the x-axis is a balance beam pivoting about a fulcrum located at the origin. The slices may be moved about and suspended from different points of this beam. The *moment* of any slice is the product of its area and its distance from the origin. If we take the slices from the sphere and the cone and suspend them both from the point $P = (-2r, 0)$, their combined moment is

$$2r[\pi(2rx - x^2) + \pi x^2] = 4\pi r^2 x \tag{9}$$

which is four times the moment of the cylinder's disk left in place at x.

Viewing our three solids as composed of infinitely many such slices, and summing (9) over all the slices, we obtain

$$2r[V_{sphere} + V_{cone}] = 4 \times \text{Moment (Cylinder)}$$

Now, as Archimedes knew, the cylinder has the same moment as if all its volume were suspended from its center of gravity i.e., the point $(r, 0)$. Therefore the cylinder has moment $rV_{cyl} = 2\pi r^4$, and so

$$2r[V_{sphere} + (8/3)\pi r^3] = 8\pi r^4$$

which can be solved to obtain

$$V_{sphere} = (4/3)\pi r^3$$

Equivalently, the volume of a sphere is 2/3 the volume of the circumscribed cylinder, a result Archimedes deemed so significant that he requested it be inscribed on his tombstone after his death.

By similar methods, Archimedes found the volumes of the ellipsoid and paraboloid of revolution, the surface areas and volumes of zones and segments of a sphere, and even the volume common to two cylinders whose axes intersect at right angles. His technique was a precursor of the integral calculus, two thousand years before Newton and Leibniz.

However, as successful as his "method of equilibrium" was as a tool for discovery, Archimedes apparently considered it lacking in rigor. Infinite processes were shunned in his day, and he probably did not really believe a volume is made up of infinitely many parallel slices. Consequently, he did not consider a formula firmly established until a rigorous, purely geometric proof was given as well.

Among the other geometric achievements of Archimedes are the formula for the area of an ellipse and a complete study of the curve now known as the Archimedean spiral (in polar coordinates, $r = a\theta$, with a constant). The area enclosed by one or more turns of this spiral was computed by means of inscribed and circumscribed circular sectors. The formula for the area of a general triangle with sides of lengths a, b, c,

$$A = [s(s-a)(s-b)(s-c)]^{\frac{1}{2}}$$

where $s = (a + b + c)/2$, although known as Heron's formula is due to Archimedes. He also solved the problem of dividing (by means of a plane section) a given sphere into two parts whose volumes were in a desired ratio. This amounted to the solution (geometrically) of a cubic equation, and Archimedes gave necessary and sufficient conditions for the existence of real roots, an extraordinary feat in the absence of our developed algebraic notation or a clear concept of function.

Although, according to the historian Plutarch, Archimedes repudiated "as sordid and ignoble the whole trade of engineering, and every sort of art that lends itself to mere use and profit," he nevertheless applied mathematical reasoning to mechanics and hydrostatics. In Plutarch's biography of the Roman consul Marcellus, we read of the catapults and other weapons of war designed by Archimedes and used by the Syracusans to ward off the attacking legions of Marcellus during the Second Punic War. Through variously arranged pulleys, levers, and hooks, Archimedes is said to have lifted whole ships in the air and dashed them against the rocks.

In more peaceful pursuits, Archimedes invented a water pump, known as the *Archimedean screw*. This consisted of a hollow tube or hose bent into the shape of a helix (coil), like a corkscrew. If one end is immersed in water, with the axis of the helix tilted at the proper angle, and if the device is rotated fast enough, water flows out of the other end. This devise was used to pump water from the holds of ships and to drain fields in Egypt after the Nile overflowed its banks. (See Exercise 4.)

In his book *On Floating Bodies*, Archimedes derived the principle that an immersed or floating object is supported or buoyed up by a force equal to the weight of the water it displaces. According to legend, Archimedes discovered this law while sitting in the public baths. Realizing the importance of his discovery, he leaped from the water and ran naked through the streets, while crying *"Eureka! Eureka!"* (I have found it).

Space does not allow us to describe the many other accomplishments of Archimedes. They can be found in Heath [19]. The following passage from Plutarch gives some indication of the awe in which this genius was held by his contemporaries and successors [11].

> It is not possible to find in all geometry more difficult and intricate questions, or more simple and lucid explanations. Some ascribe this to his natural genius; while others think that incredible effort and toil produced these, to all appearance, easy and unlabored results. No amount of investigation of yours would succeed in attaining the proof, and yet, once seen, you immediately believe you would have discovered it; by so smooth and so rapid a path he leads you to the conclusion required. And thus it ceases to be incredible that (as is commonly told of him), the charm of his familiar and domestic Siren made him forget his food and neglect his person, to that degree that when he was occasionally carried by absolute violence to bathe, or have his body anointed, he used to trace geometrical figures in the ashes of the fire, and diagrams in the oil on his body, being in a state of entire preoccupation, and, in the truest sense, divine possession with his love and delight in science.

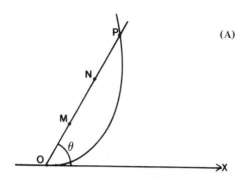
(A)

Exercises II-5

1. Let A and C be the area and circumference of a circle of radius r. Show, by means of circumscribed regular polygons, that A < rC/2 is impossible.

2. (*Angle Trisection*) Although the general angle trisection problem is incapable of solution by means of straightedge and compass alone, the problem is soluble if certain curves are admitted. One such solution involves the *Archimedean* spiral, the curve given in polar coordinates by the equation $r = a\theta$, where a is a positive constant. In figure (A), $\overrightarrow{OP}$ is an arbitrary ray making an angle θ with the polar axis $\overrightarrow{OX}$ and meeting the spiral in P. Show that circular arcs centered at O and drawn through the points M and N of trisection of $\overrightarrow{OP}$ meet the spiral in points M' and N' which are such that $\overrightarrow{OM'}$ and $\overrightarrow{ON'}$ trisect ∡POX.

3. The volume of a segment of height h in a sphere of radius a is $\pi h^2 (3a - h)/2$ [see Fig. (B)]. Suppose a sphere is to be divided into two segments of heights x and 2a - x, whose volumes are in the ratio m:n (where m and n are two given lengths). Show that

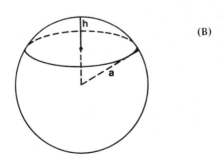
(B)

ARCHIMEDES   73

$$4a^2 : x^2 = (3a - x) : (\frac{m+n}{m})a$$

Archimedes asserted that this was a special case of the proportion $c^2 : x^2 = (b - x) : d$, or, in modern terminology, the cubic equation $x^2(b - x) = c^2 d$. He stated that a solution to this cubic is the x-coordinate of an intersection of the parabola $x^2 = c^2 y/b$ and the hyperbola $(b - x)y = bd$. He then went on to prescribe conditions under which the two curves have 0, 1, or 2 intersections, with x between 0 and b. He remarked that in the case of the spherical segment problem above a (real) solution always exists.

4. A *circular helix* is the curve described by a point moving with constant speed along an element of a cylinder while the cylinder rotates about its axis at a constant rate. Such a curve, which resembles a coil spring, is given in Cartesian coordinates by

$$x = a \cos t, \quad y = a \sin t, \quad z = bt$$

where a and b are non-zero constants. If the cylinder's axis is vertical, it can be shown that at any point of this curve, the slope of the tangent line with respect to the horizontal has the constant value b/a (see the diagrams) called the *pitch* of the helix.

Can you explain intuitively why the Archimedean screw (a helical tube wound around a solid cylinder) can draw water if the inclination of the cylinder's axis to the vertical exceeds the pitch of the helix [$\phi > \theta = \tan^{-1}(b/a)$ in the diagrams]?

5. The Greeks were eminently successful at solving first and second degree polynomial equations geometrically. We have also seen that Archimedes solved a problem involving a cubic equation. Why is

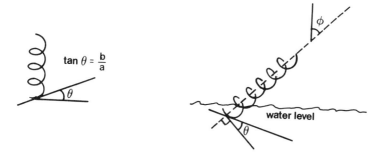

it unlikely that the Greeks ever considered problems involving equations of degree higher than three?

6. APOLLONIUS

Apollonius (260-200 B.C.) studied in Alexandria under the pupils of Euclid and went on to become a lecturer at the Museum. He was known chiefly for his eight volume work on the conic sections (*Conica*), which unlike Menaechmus he obtained as sections of a single cone by varying the inclination of the slicing plane (see Fig. II-8). If α is the cone's semi-vertex angle, and β is the angle between the cone's axis and the slicing plane, then the section is

    (i) an ellipse if $\alpha < \beta$
    (ii) a parabola if $\alpha = \beta$
    (iii) a branch of a hyperbola if $\alpha > \beta$

(Call $\beta = 0$ if the slicing plane is parallel to the axis.)

Although the conic sections had been known for some time—Menaechmus and Euclid had written about them—the *Conica* of Apollonius was so thorough and comprehensive that it superceded earlier works and became the standard reference on the subject.

Let $\overline{AB}$ be the axis of a conic section, with A a vertex. From any point P of the conic drop the perpendicular $\overline{PQ}$, which Apollonius calls an *ordinate* (Fig. II-9). Erect $\overline{AR}$ perpendicular to $\overline{AB}$ and such that the length of $\overline{AR}$ equals the latus rectum of the conic. (The *latus rectum* is the length of the chord which is perpendicular to the axis and passes through a focus. Apollonius gave a

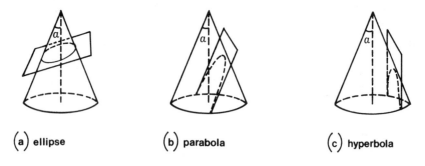

(a) ellipse   (b) parabola   (c) hyperbola

Figure II - 8.

APOLLONIUS

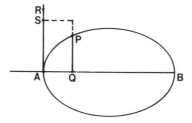

Figure II - 9.

different but equivalent definition.) Mark off on $\overline{AR}$ (or $\overline{AR}$ extended), $\overline{AS}$ so that

$$AQ \cdot AS = (PQ)^2 \qquad (10)$$

Apollonius proved that

    a. $AS < AR$ for an ellipse
    b. $AS = AR$ for a parabola
    c. $AS > AR$ for a hyperbola

The names *ellipse, parabola,* and *hyperbola,* which came from Greek words signifying deficit, comparison, and excess, respectively, originated with Apollonius and stem from these three properties.

If we denote the latus rectum AR by 4p, and set $x = AQ$, $y = PQ$, then (10) becomes $AS = y^2/x$, and the three cases become

    $y^2 < 4px$ for an ellipse
    $y^2 = 4px$ for a parabola
    $y^2 > 4px$ for a hyperbola

(In fact, the ellipse and hyperbola have equations

$$y^2 = 4px - \frac{4px^2}{d}, \quad y^2 = 4px + \frac{4px^2}{d}$$

respectively, where d is the diameter. The parabola may be thought of the limiting case as $d \to \infty$.)

Many historians see in Apollonius' methods a primitive form of analytic geometry, anticipating Descartes by eighteen centuries. To indicate the tremendous scope of the work of Apollonius, we shall describe some of the more notable results of his *Conica*.

Apollonius showed that the midpoints of a set of parallel chords of a conic all lie on one line, called a *diameter* of the conic ($\overleftrightarrow{AB}$ in Fig. II-10). The given chords are called *chords of the diameter*. The tangent line to a conic at an end point of a diameter is parallel to the chords of that diameter. For a parabola, any diameter is parallel to the parabola's axis of symmetry. For an ellipse, if C is the midpoint of diameter $\overline{AB}$ and if $\overline{DE}$ is the chord of the diameter passing through C, then $\overline{DE}$ bisects every chord parallel to $\overline{AB}$. $\overline{AB}$ and $\overline{DE}$ are called *conjugate diameters*. The chords of either diameter are parallel to the other diameter. The *axes* of a conic are conjugate diameters that are perpendicular.

Apollonius then showed that the characteristic property of a conic, (10), remains true if the axis $\overline{AB}$ in Figure II-9 is any other diameter and $\overline{PQ}$ lies on a chord of that diameter (Fig. II-10). Thus, he could transform from rectangular to oblique coordinates and vice versa.

Apollonius proved a general fact which provides a method for constructing a tangent to a parabola at a given point P. Let $\overline{PQ}$ be any chord of diameter $\overleftrightarrow{AB}$ of a parabola, and let C be the intersection of $\overleftrightarrow{AB}$ and $\overline{PQ}$ [Fig. II-11(a)]. Apollonius showed that if T on $\overleftrightarrow{AB}$ is external to the parabola and is such that AT = AC, then $\overleftrightarrow{TP}$ and $\overleftrightarrow{TQ}$ are tangents to the curve.

In addition, he proved the following result. From any point T outside a conic, let $\overline{TP}$ and $\overline{TQ}$ be the tangents drawn to the conic

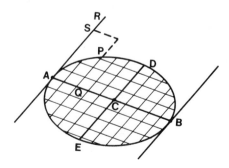

Figure II - 10.

APOLLONIUS

and let $\overleftrightarrow{TB}$ be any line through T that meets the conic, cutting it in two points A and B and intersecting $\overline{PQ}$ in C [Fig. II-11(b)]. Then

$$\frac{TA}{TB} = \frac{CA}{CB}$$

In other words, T divides $\overline{AB}$ externally in the same ratio that C divides it internally. T, A, C, B are said to form a *harmonic set* of points, line $\overleftrightarrow{PQ}$ is called the *polar* of point T and T the *pole* of line $\overleftrightarrow{PQ}$. These concepts are important in modern projective geometry.

Except for the eccentricity of a conic, which appears nowhere in the *Conica*, and the focus-directrix properties, which appear only implicitly, practically all the properties of the conic sections known today can be found in Apollonius' work.

Apollonius wrote a number of other treatises, mostly lost, but in some cases partially restored by later authors. In his two volume work *On Contacts* appeared his solution to what is now called the "Problem of Apollonius": given three things, each of which may be either a point, a line, or a circle, construct (with straightedge and compass) a circle which passes through each of the points (if any) and is tangent to each of the lines and circles (if any). For example, construct a circle tangent to three given circles. (In general, there are eight different circles tangent to the given three.)

In another work, *Plane Loci*, Apollonius proved that if A and B are fixed points and k a positive constant, then the locus of all points P satisfying

$$\frac{AP}{BP} = k$$

is a circle if $k \neq 1$ and a line if $k = 1$.

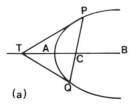
(a)

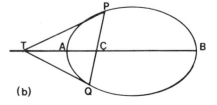
(b)

Figure II - 11.

The mathematical works of Apollonius were so complete and far reaching that many have called them the pinnacle of Greek geometry. Yet the "Great Geometer" was equally renowned as an astonomer. We shall discuss that aspect of the man in Section 7.

After Apollonius, succeeding generations of geometers did little more than expand upon earlier results of the giants. In fact it may be said that there were no major breakthroughs until the development of analytic geometry by Descartes in the sixteenth century.

Perhaps the last creative Greek geometer was Pappus (c. 320 A.D.), author of the *Collection,* a commentary and guide to the geometry of his day. There are two major theorems associated with his name; both are given in the *Collection.* One states that if A, B, and C are three points on one line, and if A', B', and C' are three points on another line, then the intersections $\overleftrightarrow{AB'} \cap \overleftrightarrow{A'B}$, $\overleftrightarrow{AC'} \cap \overleftrightarrow{A'C}$, and $\overleftrightarrow{BC'} \cap \overleftrightarrow{B'C}$ are colinear. The other theorem states that the volume generated by revolving a closed curve about an axis in the plane of the curve but not touching it is the product of the enclosed region's area by the circumference of the path described by the region's center of gravity; and the surface area of this solid is the product of the closed curve's perimeter by the circumference of the path described by the curve's center of gravity (see Exercise 4).

Many of the great geometers, including Thales, Eudoxus, Archimedes, and Apollonius were well known as astronomers. We turn next to a brief examination of Greek mathematical cosmology. (*Cosmology*—from the Greek word *kosmos,* universe—is the science or study of the universe as a whole, including its origin, history, and future evolution.)

Exercises II-6

1. Using analytic geometry, show that the midpoints of the family of chords of slope m of the parabola $y = ax^2$ lie on the line $x = m/2a$. This line is therefore a diameter of the parabola, and the chords of slope m are its chords. Show that the tangent line to the parabola at its intersection with this diameter has slope m also.

2. Using analytic geometry, verify the property of a parabola stated in the text for Figure II-11(a) for the parabola $y = ax^2$ in the case where $\overline{PQ}$ is perpendicular to the axis.

3. (a) Let A and B be two fixed points in the plane and k be a positive constant. Show analytically that the set of all points P

APOLLONIUS

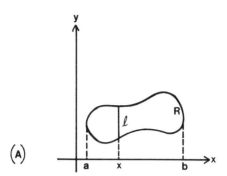

(A)

satisfying AP/BP = k is a circle if k ≠ 1 and a line if k = 1. (b) What would the corresponding loci be if A and B were two points in space?

4. (*Theorem of Pappus*)

(a) The region R in Figure A is to be revolved around the y-axis. If $\ell = \ell(x)$ is the vertical width of the region at x, then, by freshman calculus, the area of R is

$$A = \int_a^b \ell \, dx$$

and (by the "shell method") the volume generated by revolving R about the y-axis is

$$V = 2\pi \int_a^b x\ell \, dx$$

The *centroid* (or center of gravity) of R has x-coordinate

$$\bar{x} = \frac{\int_a^b x\ell \, dx}{A}$$

Show that $V = 2\pi \bar{x} A$.

(b) The length of curve C in Figure B is given by $L = \int ds$. [If C is the graph of y = f(x), then $ds = (1 + f'(x)^2)^{1/2} \, dx$ and the integral

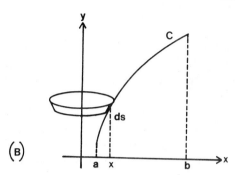

is taken from a to b.] The x-coordinate of the curve's centroid is

$$\bar{x} = \frac{\int x \, ds}{L}$$

and, by freshman calculus, the surface area obtained when C is revolved about the y-axis is

$$S = 2\pi \int x \, ds$$

Verify that $S = 2\pi \bar{x} L$. (Note: when C is a closed curve, the centroid of C is in general not the centroid of the plane region enclosed by C.)

5. Use the Theorem of Pappus (Exercise 4) to find the volume and surface area of a torus (or doughnut), obtained when a disk is revolved about a line which lies in the plane of the circle but does not touch the circle. Express your answers in terms of r, the radius of the disk, and R the distance from the disk's center to the line.

7. GREEK COSMOLOGY

Since prehistoric times man has gazed upward at the heavens in awe and wonder. Cyclical events, such as the planting of crops, changes of season, and the migrations of animals seemed linked with recurring celestial events. It was only natural that the ancient Greeks, with their philosophical and mathematical bent, should search for some underlying scheme to explain the motions of the mysterious lights that wandered in the night sky.

GREEK COSMOLOGY

The later Pythagoreans, who believed the earth was spherical, developed a theory of the universe in which the earth, the sun, the moon, and the five known planets, as well as the "sphere of the stars," all revolved around a "central fire." This central fire was hidden from us because the inhabited side of the earth always faced away from it. The earth revolved about the central fire in 24 hours, while the sun completed its circuit in 365 days.

Because 10 was considered an ideal number, the Pythagoreans believed there must be a tenth celestial object, which they called the *counter-earth*, orbiting the central fire, but on the opposite side of it from earth. We saw earlier that they believed the sizes of planetary orbits formed a progression of whole numbers, and that these moving bodies generated sounds (the "music of the spheres"). Thus, their cosmology was a mixture of geometry, number mysticism, music, and also religion.

As exploration extended the geographic frontier to regions from which the central fire and counter-earth should have been, but were not visible, these conceptions had to be abandoned. The motion of the earth about a central fire was replaced by the rotation of the earth about its axis, as was taught by Archytas, one of the last of the great Pythagoreans.

The first truly scientific theory of the heavens was that of Eudoxus. His system consisted of 27 concentric spheres, all centered at the earth. Each planet was associated with a cluster or group of four such spheres, while the sun and the moon were each associated with a group of three. A given celestial body was attached to what we might call the equator of the innermost sphere of its group, and was carried along as this sphere revolved around an axis perpendicular to the equatorial plane.

However, the axis of this sphere did not maintain a fixed direction in space. Instead, the line of this axis was attached to two diametrically opposite points of the next larger sphere of the group, and rotation of this sphere about its own separate axis caused the first sphere's axis to precess or wobble (Fig. II-12). The axis of the second sphere was linked to the third sphere in a similar manner. The outermost sphere for each celestial body rotated on an axis through the celestial poles (the line from the earth to the pole star), once every 24 hours. The axes of rotation, the rotational speeds, and the radii of the spheres were carefully chosen by Eudoxus so as to fit observations

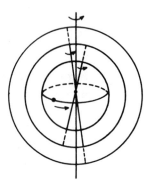

Figure II - 12.

as closely as possible. In order, from the earth outward, were the sphere of the moon, Mercury, Venus, the sun, Mars, Jupiter and Saturn. Uranus, Neptune, and Pluto were unknown to the Greeks.

Such a complicated system was necessitated by the highly irregular movements of the planets as viewed from earth. Because, as we know today, the planets revolve around the sun and not the earth, the motion of a planet would appear relatively simple only when viewed from the sun. From the earth (also in motion about the sun), the planet seems to move very erratically. In the course of its revolution about the sun, the planet will sometimes reverse direction, go backwards for a while, then reverse again and continue its forward movement. This *retrograde motion* was partially explained by the system devised by Eudoxus, although other effects, such as variations in the brightness of Venus and Mars, were unexplained.

Although the concentric spheres were merely mathematical constructs and had no real physical existence, Aristotle (384-322 B.C.) made the mistake of viewing them as material objects linked into a clockwork-like mechanism. He added 29 additional spheres to form an immense machine in which each sphere caused another one to rotate, with the whole works driven by the outermost sphere of the stars, or by some primal mover controlling the latter. Such a contrived model was eventually discredited by the scientific community, especially as more accurate observational data revealed its gross inaccuracies.

The Alexandrian astonomer, Aristarchus (c. 310-230 B.C.) was the first to impiously suggest that the sun rather than the earth is the

GREEK COSMOLOGY

center of the universe, and that the earth and planets each revolve about the sun in a circular orbit with constant speed. This was immediately attacked on a number of counts. For one thing, it did not explain why the seasons have different lengths. Moreover, it was argued, if the earth is moving around the sun, the distances to the stars (and so their apparent brightness) should vary. To this objection, Aristarchus correctly responded that the stars are so exceedingly distant that the size of the earth's orbit is insignificant by comparison. Another objection was that objects could not remain fixed to a moving earth. Also, according to the prevailing Aristotelian philosophy, all ponderable objects seek the center of the universe, and since objects fall to earth, the earth must be the center of the universe. Moreover, a force would be required to maintain the earth's motion, and no force seemed readily at hand. Finally, the retrograde motion of the planets was unexplained by the system of Aristarchus.

Aristarchus is known also for a work entitled *On the Sizes and Distances of the Sun and Moon.* In this he showed, on the basis of some rough measurements, that the distance to the sun is about 19 times the distance to the moon (it is actually about 389 tmes as distant), and that the sun is 6 to 7 times larger than the earth in diameter (actually about 109 times). Such large errors were due to the crudeness of his instruments and the lack of a good enough estimate for π. Trigonometry had not been invented yet. Nevertheless, he was the first to investigate scientifically the scale of distance within our solar system: others had considered only describing and predicting angular directions of objects in the sky.

Quantitative astronomy became firmly established with Apollonius who was the outstanding astronomer of the third century B.C. Much of his work dealt with irregularities in the motion of the moon. He is believed also to have been familiar with the theory of epicycles that became the basis of the Ptolemaic theory, discussed below.

The culmination of Greek astronomy was the work of Hipparchus (died c. 125 B.C.) and Claudius Ptolemy (2nd century A.D.). Before we can fully appreciate their contribution, we need to know a few facts of astronomy.

In the following discussion, we shall often want to compare how planetary orbits appear from the *geocentric* point of view (in which the earth is regarded as at rest at the center of the universe) with how they appear from the *heliocentric* point of view (in which

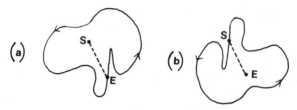

Figure II - 13.

the sun is taken as the stationary center of the universe). It is a simple matter to transform a diagram depicting one of these hypotheses into a diagram depicting the other. For example, imagine that the sun, S, is stationary and that the earth, E, is in motion about S along the highly unusual orbit pictured in Figure II-13(a). Figure II-13(b) represents how the same situation would look (at the very same moment) from the point of view of an observer on E, who considers his world to be at rest, with S in motion around it. Notice that the distance and direction from S to E is the same in both figures. We can transform either picture into the other by rotating 180° and then switching the labels S and E. The same would be true for an orbit of any shape.

Let us now look at the actual orbit of the earth. According to modern (Newtonian) physics, the earth follows an elliptical path around the sun, with the sun at one focus as shown in Figure II-14. The eccentricity, actually about 0.017, has been exaggerated: the orbit is more nearly circular. Perihelion and aphelion are, respectively, the points of closest and farthest distance from the sun.

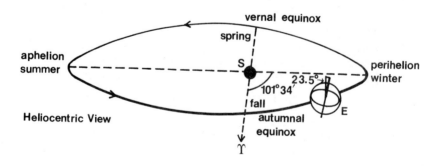

Figure II - 14.

The earth's polar axis is not perpendicular to the plane of its orbit, but is inclined about $23\frac{1}{2}°$ away from the plane's normal direction. This angle is responsible for the seasons, since the northern hemisphere is tilted toward the sun at aphelion (summer) and away from the sun at perihelion (winter). (Of course, the seasons are reversed in the southern hemisphere.) It may seem strange that the sun is actually farther away in summer than in winter, but remember the orbital eccentricity is small, so that the variation in earth-to-sun distance produces only a slight effect.

The other features of Figure II-14 are more easily explained if we now transform to the corresponding geocentric diagram, Figure II-15. Here the sun follows an elliptical path about the earth (although the Greeks sought to describe this path, as well as the orbits of the planets, in terms of circles or spheres). The ancients called the apparent path of the sun across the "celestial sphere" the *ecliptic*, because lunar eclipses occur only when the moon is on or near this path. (To the ancients, the star-studded night sky resembled an upturned bowl centered at the earth. Even today, the astronomer who is concerned only with the directions of the stars and not their distances thinks of the stars as distributed over a spherical surface, the celestial sphere.)

Because of the inclination of the earth's polar axis described above, the plane of the ecliptic makes a $23\frac{1}{2}°$ angle with the earth's equatorial plane. The intersection of these two planes (in the geocentric picture, a line through the earth's center) is called the *line of nodes* and points to the two locations on the ecliptic when the sun is

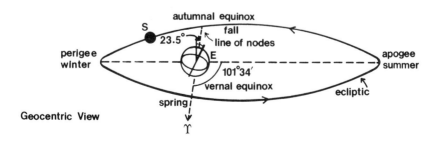

Figure II - 15.

directly overhead at midday, and at every point on earth, night and day are equally long. These are the vernal and autumnal *equinoxes,* occurring about March 21 and September 22, respectively.

At the time of the vernal equinox, when each year the sun first crosses the equatorial plane, the line of nodes, in the direction from the earth to the sun, meets the celestial sphere in a point known as the *First Point of Aries.* Astronomers label this point ♈, the traditional symbol for the constellation Aries (the ram), the first of the twelve signs of the zodiac.

Now it turns out that the earth's axis of rotation (or polar axis) does not maintain a fixed direction in space. Instead the axis slowly precesses, much as the spin axis of a top precesses about the vertical direction. The precessing axis (from the geocentric point of view) sweeps out a cone with vertex at the earth's center and semi-vertex angle of $23\frac{1}{2}°$, completing one full sweep in about 26,000 years. This in turn, causes the line of nodes to turn at the rate of 360° in 26,000 years, or nearly $1\frac{1}{2}°$ per century. (In 13,000 years, summer and winter in Figure II-15 will be interchanged.) Because of this *precession of the equinoxes,* as the rotation of the line of nodes is called, the *First Point of Aries* is not presently in the constellation Aries as it was in ancient times, but is now in the constellation Pisces. In another two millenia or so, the line of nodes will point to the constellation Aquarius. Figures II-14 and II-15, with 101° 34' as the angle between the line of nodes and the ellipse's axis, represent the situation in 1920, often taken as a standard reference year.

Now consider the case of the earth (E) and another planet (P), both orbiting the sun, as in Figure II-16(a), where for simplicity the orbits are drawn as perfect circles. Figure II-16(b) shows the corresponding geocentric view, where S is considered to orbit E. Since the sun-to-planet distance must be constant, P revolves on a circle whose center is the moving sun S. [Note that at corresponding instants, $\triangle$EPS in Fig. II-16(a) is congruent to its counterpart in Fig. II-16(b).]

Some time during the period between the time of Eudoxus and the time of Apollonius, the Greeks evolved an astronomical theory in which planetary motion was described as in Figure II-16(b). In Figure II-17, where a planet P and its orbit are shown at several

GREEK COSMOLOGY

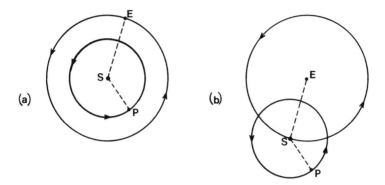

Figure II - 16.

instances, we see how the planet, when viewed from the earth, appears to reverse its direction periodically. This is the retrograde motion described earlier. The moving circle centered at S is called an *epicycle* and the circle centered at E (along which S moves) is called a *deferent*.

Actually, in the case of an outer planet, like Jupiter or Saturn, the picture is somewhat different, as we can see by making the transformation from the heliocentric [Fig. II-18(a)] to the geocentric point of view [Fig. II-18(b)], the latter now yielding what is called an *eccentric circle* picture. However, the diagram of Figure II-18(b) can be made to look like an epicycle-deferent picture by the

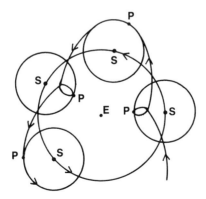

Figure II - 17.

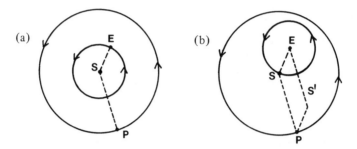

Figure II - 18.

following device. Complete the parallelogram two of whose sides are $\overline{SE}$ and $\overline{SP}$. This determines a point S' which describes a circle centered at E (since ES' = SP = constant). Moreover S'P = ES is constant. Hence P rides on a circle (epicycle) centered at S' which in turn rides on a circle (deferent) centered at E. In this case the center of the epicycle is just a mathematical point and not a material body.

The epicyclic theory achieved its greatest elaboration in the hands of Hipparchus, the most eminent Greek astronomer. He is generally credited with the invention of trigonometry and was the first to employ longitude and latitude in measurements on the earth.

Hipparchus was compulsively precise and made extremely careful astronomical measurements over most of his lifetime. He also had access to Babylonian records of observation, some dating back many centuries. Hipparchus found that the point on the celestial sphere about which the stars seem to revolve in the course of a night had changed over the centuries. In other words, he had discovered the precession of the earth's axis! He estimated the resulting precession of the equinoxes at about 59" per year (it was actually about 50.2").

He made a number of other measurements that were surprisingly accurate for his time. He determined the length of a year to within about $6\frac{1}{2}$ minutes, measured the departure from circularity of the earth's orbit (essentially what we would call its eccentricity today), measured the periods of the five known planets (as viewed from the earth), and charted irregularities in the motion of the moon. He could predict a lunar eclipse to within just a few hours.

To make theory fit his observations more closely, Hipparchus suggested higher-order epicycles. In this variant of the theory, the

GREEK COSMOLOGY

planet P moves on a third circle whose center rides on an epicycle centered at the sun. In order to improve the model, succeeding astronomers had to compound the order of epicyclicity even further, but little real theoretical progress was made until the rise of the heliocentric theory between the time of Copernicus (1473-1543) and Kepler (1571-1630) over 1600 years later.

The source of most of our knowledge of the work of Hipparchus is the monumental book of Claudius Ptolemy (died c. 168 A.D.), known as the *Almagest* (from an Arabic title derived from the Greek *megiste syntaxis*, or "greatest collection"). The Almagest contains the foundations of trigonometry (plane and spherical), and the details of the epicyclic theory worked out by Hipparchus. The value of π used was

$$\pi = 3° \ 8' \ 30'' \approx 3.14167$$

(The Babylonian sexigesimal system found its way into the reckoning of angles and survives to this day.)

Though based largely on the writing of Hipparchus, the style and completeness of the *Almagest* quickly made it the standard work on astronomy, and it remained so until the time of Copernicus. Indeed, the description of the heavens detailed in this work has become known as the *Ptolemaic theory*.

Exercises II-7

1. Eratosthenes (c. 230 B.C.) made a famous measurement of the earth. He observed that at the summer solstice, at Syene (present day Aswan), a vertical stick cast no shadow, while at Alexandria, on the same meridian as Syene, the sun's rays were inclined $1/50^{th}$ of a complete circle to the vertical ($\theta = 2\pi/50$ radians, in the figure on p. 90). He then calculated the circumference of the earth from the known distance of 5000 stadia from Alexandria to Syene. If a stadium is 516.7 feet, what was Eratosthenes' estimate, in miles? (The actual mean circumference of the earth is 24,874 miles.)

2. According to the theory proposed by Nicholas Copernicus, the sun is the center of the universe, and the motions of the moon and planets follow an epicyclic scheme similar to that of the Ptolemaic theory, but requiring fewer circles. In the figure on p. 90, P is a planet moving on an epicycle whose center C (a mathematical point)

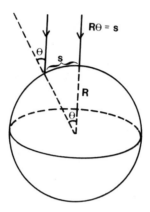

revolves about S, the sun. Let Ω be the angular speed of C about S, and let ω be the angular speed of P about C. Show (by means of a sketch) that if the ratio ω/Ω is suitably chosen, the path of P will be an elongated circle, resembling the actual elliptical orbit. What is a suitable value for the ratio?

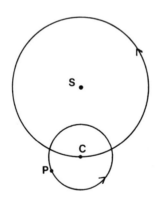

III

THE AXIOMATIC METHOD

> *One must be able to say at all times—instead of points, straight lines, and planes—tables, chairs, and beer mugs.*
> —David Hilbert

> *Mathematics may be defined as the subject in which we never know what we are talking about, nor whether what we are saying is true.*
> —Bertrand Russell

1. PIPS AND GLOBS

The vehicle employed by Euclid in his *Elements* was the method of deductive logic introduced by the early Greek geometers 2500 years ago. It has been the mathematician's principal tool down to the present day. Before we can proceed to the controversy surrounding Euclid's parallel postulate and to the study of alternative and more general geometries, it is necessary first to make clear the nature of the axiomatic method, both as used by the Greeks and as it is understood today.

Each proposition or *theorem* in a deductive argument must be logically derived from earlier propositions which in turn must be deduced from still earlier propositions, and so on. Since this process cannot retreat into the past indefinitely, the early Greek geometers realized that there must be some primary initial statements which we do not attempt to prove, but whose truth is readily acceptable to the reader, and from which the process of deduction begins. These are the *postulates* or *axioms* (words which are interchangeable in current usage). See Appendix A for the postulates chosen by Euclid.

Since the statements of our discourse, including the postulates themselves, will undoubtedly contain terms peculiar to the study at hand (triangle, dihedral angle, parallelogram, etc.), and since these must be defined, the Greeks also saw the need to specify an initial collection of basic terms (point, line, plane, angle, etc.) whose meanings either were assumed clear or suggested to the reader, and in terms of which other technical terms might be defined as needed. The following examples will illustrate these ideas.

Example 1a

A set S of *trees* is arranged in such a way that certain subsets of S are arranged in *rows*. (The basic terms are therefore "tree" and "row," which is a collection of trees. Assume that S and the rows are non-empty and finite.) The following rules (postulates) are stipulated.

P1. Each tree of S belongs to at least one row.
P2. Any two distinct trees of S belong to a unique row.

Before we can state the third postulate, we need a definition.

Definition III-1a

A given row is called *separate* from another given row if these two rows have no tree in common.

P3. For each row, there is one and only one row separate from it.

From the information given in these three postulates, we may now derive (deduce) the following consequences.

Theorem III-1a

Every tree belongs to at least two rows.

Proof: Let t be any tree of S. Then t belongs to some row A, by P1. By P3, there is a unique row B separate from A. Let u be a tree in B. Since t ≠ u, P2 tells us that there is a third row C to which t and u

PIPS AND GLOBS 93

both belong. (C is distinct from A and from B because C contains u, which is not in A, and t, which is not in B.) Hence t belongs to at least two different rows (A and C). Since t was arbitrarily chosen, the same holds for every tree in S.

Theorem III-2a

Every row contains at least two trees.

Proof: Let A be any row, and let t be a tree in A. By Theorem III-1a, t belongs also to a second row, B. Using P3, let C be the row separate from B (Fig. III-1). If A contained no tree other than t, then C would be separate from A, as well as from B. This would contradict the uniqueness asserted in P3. Hence A must contain at least one other tree. Since the choice of A was arbitrary, we have shown that every row contains at least two trees.

From Theorem III-2a and the postulates, the reader should be able to prove the next theorem.

Theorem III-3a

There exist at least six rows.

The next theorem is more difficult to prove. (See Exercise 6.)

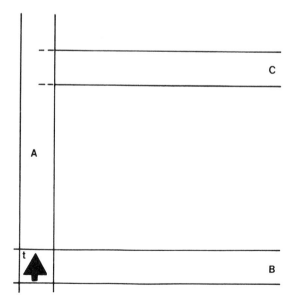

Figure III - 1.

Theorem III-4a

Each row contains exactly two trees. Using Theorem III-4a, we can deduce the following.

Corollary

There are in S exactly four trees and exactly six rows.

Proof: Beginning with two separate rows containing two trees each, we have four trees. These determine six rows, six being the number of ways that a pair can be selected from a set of four elements. It is easy to verify that these are all distinct, and that the existence of additional trees would contradict the postulates.

Example 1b

We are given a polyhedron consisting of a (finite, non-empty) set of *vertices* and certain (non-empty) collections of vertices called *edges*. The following postulates apply.

P1. Every vertex lies on at least one edge.
P2. Any two distinct vertices lie on a unique edge.

Definition III-1b

A given edge is called *divergent* from another given edge if no vertex lies on both of them.

P3. For every edge, there is one and only one edge divergent from it.

We can now deduce the following theorems.

Theorem III-1b

Every vertex lies on at least two edges.

Theorem III-2b

On every edge lie at least two vertices.

Theorem III-3b

There exist at least six edges.

Theorem III-4b

On each edge lie exactly two vertices.

Corollary

There are altogether exactly four vertices and exactly six edges.

However, it would be a needless duplication of effort to write down proofs of these theorems. If we compare the above with Example 1a, we see that under the correspondence

> tree ↔ vertex
> row ↔ edge
> belongs to ↔ lies on

they are formally the same. Consequently, if we use the above "dictionary" to translate statements about trees and rows into statements about vertices and edges, we obtain for Example 1b six theorems which are the analogues of the six theorems of Example 1a (the proofs translate also!).

In his *Elements*, Euclid defines a point as "that which has no part," and a line as "breadthless length." Nowhere does he define what length and breadth are. From a logical standpoint, these "definitions" are useless, and, in fact, are not used by Euclid in his proofs. They are probably intended to recall to the student familiar mental images of these idealized concepts, and to insure that the teacher and pupil are speaking of the same things. Many of the other definitions are logically unsound. For example, we are told that "a plane angle is the inclination one to another of two lines in a plane," but the meaning of inclination is not specified. A worse example is Euclid's definition of ratio, given in Section II-3).

In the modern view, the idea of initially defined terms is rejected, since it may be argued that the initial terms depend for their meaning on still more basic terms, and then these terms need to be defined, etc. Just as we cannot prove everything, but must leave some statements (the postulates) unproved, so too we cannot define every one of our terms, but must begin by leaving certain terms *undefined.* All that we need know concerning these undefined entities and their relationships is to be found in the postulates. (In this sense then, postulates are really definitions in disguise!)

Example 2

We are given a set of objects called *pips*. Certain collections of pips are called *globs*. These pips and globs satisfy the following postulates.

P1. There exist at least two pips.
P2. Given two distinct pips, there is one and only one glob containing them both.
P3. Given any glob, there is a pip not in it.
P4. Given any glob G and any pip p not in G, there exists one and only one glob G' such that p is in G', and G and G' are disjoint (i.e., have no pips in common).

P1 is included to prevent the set of pips and the set of globs from being empty. For convenience, if a pip p belongs to two globs G and H, we shall say that G *intersects* or *meets* H in p, or that G and H *pass through* p. The unique glob containing two pips p and q will be denoted pq.

Without the faintest idea what pips and globs might be, we nevertheless can deduce the theorems below from the information contained in the postulates. In the sketches below, a pip is represented by X and a glob by the plane region within a closed curve. The sketches are only mental aids, and we could just as well represent pips as trees and globs as rows, or we could have used any other convenient scheme suggested by the postulates.

Theorem III-5
   Two distinct globs have at most one pip in common.
   Proof: This is immediate from the uniqueness asserted in P2.

Theorem III-6
   Two distinct globs each disjoint from a third glob are disjoint from each other.
   Proof: Let distinct globs G and H each be disjoint from glob K. If G and H had a pip p in common, then there would be two globs disjoint from K and containing the pip p, which is not in K. This contradicts P4. Therefore, G and H must be disjoint. (Note that Theorem III-6 rules out the configuration shown in Fig. III-2.)

Theorem III-7
   Every glob contains at least two distinct pips. First we prove an intermediate result.

Lemma III-8
   Every glob contains at least one pip.

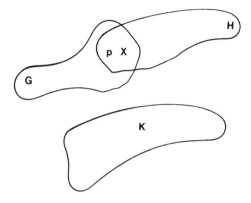

Figure III - 2.

Proof: Suppose, on the contrary, that there exists an empty or "pipless" glob, G. By P3 and P4, there is a pip p contained in a glob H disjoint from G (Fig. III-3). By P3 again, there is a pip q outside of H. Let K = pq. Since G has no pips at all, H and K are both disjoint from G. However, they are not disjoint from each other, and this contradicts Theorem III-6. Accordingly, there can be no pipless globs. Every glob contains at least one pip.

Proof of Theorem III-7. Assume, on the contrary, that there is a "unipip" glob, G. Apply P3, P2, P3 and P4, in that order, to deduce a contradiction to Theorem III-6.

With regard to the following theorems, recall that two sets (whether finite or infinite) are said to have the same number of

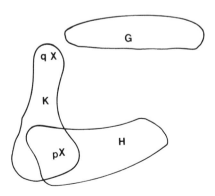

Figure III - 3.

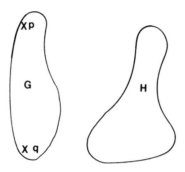

Figure III - 4.

elements if there is a one-to-one correspondence between their elements.

Theorem III-9

The number of globs containing a given pip is the same as the number of globs containing any other pip.

Proof: Let p and q be any two pips, and let G = pq. By P3 and P4, there is a glob H disjoint from G (Fig. III-4). Each glob through p other than G itself will meet H in a unique pip (why?). This furnishes a one-to-one correspondence between the globs through p that meet H and the pips of H. Similarly, there is a one-to-one correspondence between the globs through q that meet H and the pips of H. Remembering to include G itself, we find that the same number of globs pass through p as through q.

Theorem III-10

The number of pips contained in a given glob is the same as the number of pips contained in any other glob.

Proof: If G is any glob, let p be a pip not in G (by P3). Except for the unique glob containing p and disjoint from G, every glob through p has a unique pip in common with G, with different globs determining different pips and vice versa. Consequently, the number of pips in any glob is one less than the number of globs through any pip.

Together with Theorem III-7, this yields the following.

Corollary III-11
  Every pip belongs to at least 3 distinct globs.

  The preceeding postulate system will play no further role in this book. It was introduced solely because it illustrates the idea of *undefined* or *primitive* terms in a logical discourse. It is important to realize that the theorems were deduced from the postulates without any meaning or interpretation assigned to the words "pip" and "glob." Realization of this fact is crucial to an understanding of non-Euclidean geometries. In the course of our investigations, you will encounter postulates and theorems which will undoubtedly appear very strange. For example, Lobachevsky's Postulate stipulates the existence of *more* than one line parallel to a given line through a given external point. Before protesting the impossibility of such a thing happening, you should remember that "point" and "line" are undefined terms (just as were "pip" and "glob"), which are not necessarily—and in fact cannot be, in Lobachevskian geometry—interpreted as meaning ordinary points and lines of the flat Euclidean plane. Whenever you feel troubled by a strange-sounding statement, it might be a good idea for you to replace the words "point" and "line" by "pip" and "glob," or other nonsense words.

  In the modern view, undefined terms are considered much like variables, and the postulates and theorems are not statements at all but "statement forms." Just as an equation, e.g., $3x^2 - 3y + z^2 = 0$, may or may not hold depending on what values are assigned to the variables, so too the truth or falsity of a statement form depends on the interpretations we put on the undefined terms.

  (Of course, the truth or falsity of the assertion that a particular theorem is deducible from the postulates does *not* depend upon a choice of interpretation. For example, it is true that Theorem III-9 is implied by the postulates of Example 2. The derivation depended only upon an application of the rules of deductive logic to the postulates and had nothing to do with any particular interpretation of the words "pip" and "glob.")

  If the interpretations placed on the undefined terms convert the postulates into true statements, then the theorems (assuming the derivations are correct) become true statements also. Such an interpretation is called a *model* for the postulate system. Here are some models for the previous system of postulates.

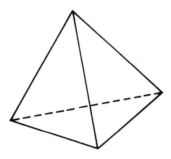

Figure III - 5.

i. The pips are the four vertices of a tetrahedron. The globs are the six edges. In Figure III-5 an edge is viewed as (determined by) a set of two vertices. "Pip *belongs to* glob" is interpreted as "vertex *lies on* (is an end point of) an edge."
ii. The pips are the points labeled A, B, C, and D in Figure III-6, while the globs are the two-point sets AB, AC, AD, BC, BD and CD, which are represented by lines in the figure.

Definition III-12

An *isomorphism* between two models of a given set of postulates is a one-to-one correspondence between the elements, relations and operations (respectively) of one model and those of the other, such that whenever a given relationship holds among certain elements in one model, the corresponding relationship holds among the corresponding elements in the other.

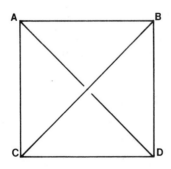

Figure III - 6.

PIPS AND GLOBS

For example, the logarithm function establishes an isomorphism between the group of positive real numbers under the operation of multiplication and the group of all real numbers under the operation of addition: x corresponds to log x; multiplication in the first group corresponds to addition in the second. If z = x × y, then log z = log x + log y.

It is obvious that models (i) and (ii) are formally the same, and in fact isomorphic. Figure III-6 is just a flattened tetrahedron. An isomorphism between these two models is obtained by labeling the vertices of the tetrahedron A, B, C and D (in any one of 4! = 24 ways).

  iii. Pips and globs are, respectively, points and lines of the Euclidean plane. In this interpretation, P4 is the statement that one and only one line parallel to a given line may be drawn through a given external point.
  iv. (*The Rational Plane*) Pips are points with rational coordinates in the *Cartesian plane* (i.e., the Euclidean plane with rectangular coordinates). A glob is the set of all pips satisfying an equation of the form ax + by + c = 0, with a, b, and c rational numbers and with a and b not both zero. (See Exercise 7.)
   v. The pips are the nine points labeled A through I in Figure III-7. The globs are the three-point subsets lying on the eight straight lines drawn in the figure, plus the subsets AHF, CDH, GBF, and IBD-12 globs in all. (Verify the postulates.)

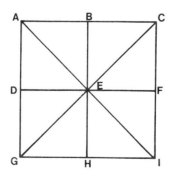

Figure III - 7.

Any theorem, once deduced from the postulates, is a true statement in each and every model for these postulates: a great economy of thought. Moreover, by working formally, with the postulates, we focus on the underlying structure common to all the models, and free ourselves from irrelevant or inessential features that may be peculiar to a particular model. This is one advantage of the axiomatic method.

In the natural sciences, general laws, such as Newton's law of gravitation, are inferred from observation of particular instances (the falling of an apple, the orbiting of satellites and planets, the paths of comets). However, we can never be absolutely certain that a physical law is valid in all instances and under all possible conditions. Indeed, modern physics has established that Newton's law is only an approximation valid under certain conditions. For very high speeds, very great masses, or vast distances, better (more accurate) laws of gravitation are to be found in Einstein's theory of general relativity.

In contrast to the "relative" truth of physical laws, a correct mathematical deduction of a theorem from a set of postulates possesses truth of an absolute kind. Assuming the deduction is logically correct, the theorem will be true in every model that is now known or will ever be discovered for these postulates.

We give one final example illustrating that relational phrases such as "belongs to," "lies on," "is between," etc., are also undefined terms and are subject to interpretation.

A binary relation R on a set S is a set of ordered pairs of elements of S. If (a,b) belongs to R, we write aRb.

Example 3

We are given a set S of (undefined) elements and an (undefined) binary relation R on S, such that for any a, b, c in S,

P1. If aRb, then a and b are distinct.
P2. If aRb and bRc, then aRc.

As models, we can give

i. S = {positive integers}, R is < ("less than"), i.e., R is the set of all ordered pairs (a,b) of positive integers for which a < b.
ii. S = {people in the world}, R is "is an ancestor of."

iii. S = {plane polygonal figures}, R is "has smaller area than."
iv. S = {cardboard boxes}, R is "fits inside."

Exercises III-1

1. Consider the following postulate set, in which *pea* and *pod* are the primitive (undefined) terms.

   P1. Every pod is a collection of peas.
   P2. Any two distinct pods have exactly one pea in common.
   P3. Every pea belongs to exactly two pods.
   P4. There exist exactly four pods.

(a) Deduce the following theorems from these postulates.

Theorem 1. There are exactly six peas.
Theorem 2. There are exactly three peas in each pod.
Theorem 3. For each pea, there is exactly one other pea not in any pod with it.

(b) Give a model for this postulate set.
2. Supply the details for the last sentence in the proof of the Corollary to Theorem III-4a.
3. Consider the following postulate system, in which the undefined objects are *abbas* and *dabbas*, and *glops* is an undefined relation which may hold between a dabba and an abba.

   P1. There exists at least one abba.
   P2. Each dabba glops at least one abba.
   P3. Each abba is glopped by at least one dabba.
   P4. Any two distinct dabbas glop a unique abba.

Definition. Two distinct abbas are called *shmerk* if there is no dabba which glops them both.

   P5. For each abba, there is one and only one other abba such that the two are shmerk.

Which of the following interpretations are models for this postulate set? (For each that is not a model, state which postulates fail.) In each interpretation, what does *shmerk* mean?

a. dabba = a point of the Euclidean plane
   abba = a line of the Euclidean plane
   glops = 'lies on' (so that 'glopped by' = 'contains')
b. dabba = a great circle on the sphere
   abba = a point of the sphere
   glops = 'contains' (so that 'glopped by' = 'lies on')
c. dabba = a vertex of a given tetrahedron
   abba = an edge of the tetrahedron
   glops = 'is an end point of' (so that 'glopped by' = 'contains as endpoint')

4. Deduce the following theorems from the abba-dabba postulates of the previous exercise.

Theorem a. Every dabba glops at least two abbas.
Theorem b. Every abba is glopped by at least two dabbas.

5. To the ancient Greeks, an axiom was not only true, but self-evident. In the modern view, an axiom is neither true nor false. Explain.

6. Deduce the following theorems from the postulates of Example 1a. (You may use Theorems III-1a, III-2a, and III-3a.)

Theorem a. The number of rows containing a given tree is the same as the number of rows containing any other tree.
Theorem b. The number of trees contained in a given row is the same as the number of trees contained in any other row.

(Note: With these theorems established, we can now prove Theorem III-4a, as follows. All rows contain the same number of trees, n. Suppose $n \geq 3$. We shall deduce a contradiction. Let A and B be two distinct rows containing trees $a_1, a_2, \ldots, a_n$ and $b_1, b_2, \ldots, b_n$, respectively. Let C be the unique row containing a_1 and b_1. Then there are n-1 distinct rows each containing a_2 and a tree of B other than b_1, and there are n-1 rows each containing a_3 and a tree of B other than b_1. Only one of these two collections of n-1 rows can contain the row separate from C. Consider the other collection. Each

PIPS AND GLOBS

of its rows must meet C in a unique tree. These n-1 trees are distinct from one another and must be distinct from a_1 and b_1. Thus C contains at least n+1 trees, an impossibility). (This proof was contributed by Joseph Karpicz.)

7. Prove that P4 of Example 2 holds in the rational plane. (This is not entirely trivial. What if the solution set of a pair of glob equations were an irrational point? Can that happen?)

2. PROPERTIES OF AXIOM SYSTEMS

Consistency

A system of postulates is called *model consistent,* or simply, *consistent* if there exists a model for the system. This implies that it is not possible to deduce from the system a theorem which contradicts a postulate or a previously deduced theorem. Thus the usual Cartesian coordinate model (Hilbert [21], Chapter II; Pogorelov [35], Chapter III) establishes the consistency of Euclidean geometry, provided we agree that arithmetic is consistent. (In like manner, essentially every proof of a postulate system's consistency ultimately rests on the consistency of some other system. Consequently, the use of models proves only what might be called "relative consistency.") Beltrami (see Section VI-3) established the consistency of non-Euclidean geometry in 1868 by giving a model for it. Consistency is a property that is required of any postulate system. We shall assume consistency in the following definitions.

Independence

A postulate is called *independent* if it is not a logical consequence of the other postulates of the system under discussion. How does one verify independence? Well, if a postulate P *were* a logical consequence of the other postulates, then P would necessarily hold in every model for these other postulates. Therefore, to show P is independent, one would have to establish the existence of a model in which the other postulates hold but P does not. For example, interpreting the trees and rows of Example 1a as points and lines in the Euclidean plane shows that P3 is independent.

In order to demonstrate that an entire set of postulates is independent, we must, for each given postulate, give a model in which all the other postulates, plus the negation of the given one, hold.

Independence is not a required property for a postulate system, although it is often desired. Mathematicians often find it esthetically pleasing to have a theory based on as few initial assumptions as possible. On the other hand, strict insistence on independence sometimes results in an extraordinary amount of work to establish theorems which might be almost immediate consequences of a slightly redundant postulate set. The postulates for a group that appear in Exercise 4 are independent, but it requires some tricky reasoning to deduce from them even the existence of a two-sided identity element.

Completeness

A system of postulates is *complete* if it is impossible to add to it a new independent postulate which is consistent with the given set, and which does not contain any new terms. Equivalently, any statement expressible in the terms of the system is either provable or disprovable (i.e., its negation is provable).

Euclid's system of postulates was incomplete, as we shall see shortly. Hilbert's system without his parallel postulate is incomplete. Completeness is demonstrated by showing categoricity, which implies completeness.

Categoricity

A set of postulates is *categorical* if there exists an isomorphism between any two models of the set. As stated in Definition III-12, this means that there exists a one-to-one correspondence between the elements, relations and operations of one model and the other such that whenever a given relationship holds among elements in one model, the corresponding relationship holds among the corresponding elements in the other.

It is easy to see why a categorical postulate set S must be complete. If S were not complete, then there would exist an independent statement P which is consistent with the set S. Therefore there is a model for S and P. On the other hand, because P is independent of S, it can be shown that there is another model for S in which P is false. Since these two models cannot be isomorphic, S could not be categorical (but we assumed it is). This contradiction implies that S must be complete.

It turns out that completeness does not imply categoricity.

PROPERTIES OF AXIOM SYSTEMS
107

Examples

  i. The axioms for a *field* are not categorical. The field of real numbers is not isomorphic to the field of rational numbers, nor to the finite field of integers modulo a given prime.

  ii. The postulate set of Example 1a is categorical. Any model is isomorphic to the tetrahedron model. (The categoricity is a consequence of the corollary to Theorem III-4a.)

  iii. Hilbert's postulates for Euclidean geometry (Appendix B) are known to be categorical.

  iv. The postulate set of Example 3 is consistent and independent, but not categorical.

  v. In modern algebra, the axioms for "ordered integral domain in which the positive elements are well-ordered" are categorical (McCoy, [30, p. 63]).

  vi. Definition: A binary relation R on a set S is called an *equivalence relation* if, for every a, b, and c in S,

(1) aRa,   (2) aRb implies bRa,   (3) aRb and bRc imply aRc

Show that these three properties are consistent and independent.

Exercises III-2

1. Prove the independence of P2, P3, and P4 in the postulate set of Exercise III-1, 1.

2. Let K be a set of undefined elements, and let R be an undefined binary relation on K. Assume that for any a, b, and c of K:

P1. If a and b are distinct, then aRb or bRa (or both).
P2. If aRb, then a and b are distinct.
P3. If aRb and bRc, then aRc.

(a) Prove this postulate set is consistent. (b) Deduce: Theorem. If aRb, then it is not true that bRa. (c) Prove this postulate set is independent.

3. Let S be a set of undefined elements, and let R be an undefined binary relation on S. Assume that for any a, b, and c in S:

P1. If aRb, then a and b are distinct.
P2. If aRb and bRc, then aRc.

P3. For each a, there exist elements x and y in S such that xRa and aRy.

(a) Prove this postulate set is consistent. (b) Prove this postulate set is independent.

4. Definition. A *group* is a set G together with a binary operation on G, ∘, such that

P1. ∘ is associative, i.e., $(x \circ y) \circ z = x \circ (y \circ z)$ for all x, y, and z in G.
P2. For any a, b in G, there exist elements x and y in G such that $a \circ x = b$ and $y \circ a = b$.

(Note: a *binary operation* ∘ on G is a rule assigning to each ordered pair of elements a, b in G an element a∘b.)

(a) Use the following interpretations to prove these postulates are independent:

i. Let G_1 = IR (the set of all real numbers), and define $a \circ b = (a + b)/2$, for all a, b in G_1.
ii. Let G_2 = IR and define $a \circ b = a$, for all a, b in G_2.

(b) Give a different pair of interpretations that prove independence. (It can be shown that P1 and P2 imply the existence of a two-sided identity element in G and the existence of a two-sided inverse for each element of G.)

3. EUCLID AND THE FOUNDATIONS OF GEOMETRY

Throughout his *Elements,* Euclid makes a number of tacit assumptions that are not deducible from his stated postulates (given in Appendix A). In many cases, he relies on relationships suggested by his diagrams but not proved. We are not attacking the validity of any of these tacit assumptions, but rather the fact that they are not made explicit. The point is that any assertions that are used, if not themselves postulates, must be logically deduced from the postulates or from previously deduced theorems or propositions. This is a prime rule of the axiomatic method.

EUCLID AND THE FOUNDATIONS OF GEOMETRY

A. Superposition

In proving his Proposition I,4 (the "side-angle-side theorem" or SAS—see Appendix A), Euclid places ("applies") one triangle upon another. He seems to be saying that figures may be moved about without distortion. Such an idea is not contained in or even implied by his postulates. We shall learn later that, in the application of geometry to physical space, the possibility of motion without distortion (i.e., the existence of "rigid motions") is actually an assumption about the curvature of space (Chapter VIII). Such motion is impossible on the surface of an egg or a pear, for example.

Euclid used the method of superposition infrequently and with apparent reluctance; but he evidently could find no other way to prove the SAS proposition. Since the concept of motion or transformation is missing from his explicit assumptions, his proof of this theorem is invalid. Many modern postulate systems for Euclidean geometry, including those of Hilbert (Appendix B), Birkhoff [4], and the School Mathematics Study Group [41], take SAS as a postulate. Some other axiom systems include "motion" among their primitive terms (e.g., Pogorelov, [35]).

B. Continuity

In many of his constructions, such as copying segments and dropping or erecting perpendiculars, Euclid alludes to points whose existence cannot be proved from his postulates.* Such points are the intersections of a line and a circle or of two circles.

To see that the existence of such points is independent of his postulates, consider (as a model for Euclid's five postulates) the rational plane (model iv for the pips and globs postulate set). Here the line with equation $y = x$ does not have a *rational* point in common with the circle whose equation is $x^2 + y^2 = 1$, even though this line joins the interior point $(0,0)$ to the exterior point $(1,1)$.

Consequently, to guarantee the existence of such points of intersection, some type of continuity axiom is needed. (This will ensure that lines and circles have no "holes.") In 1872, Dedekind formulated the following postulate.

---
*He also assumes, often without justification, particular locations for these points.

> If all points of a straight line fall into two classes such that every point of the first class lies to the left of every point of the second class, there exists one and only one point which produces this division of all the points into two classes, this separation of the straight line into two parts.

Instead of using the terms "left" and "right" (which would be meaningless for a vertical line), we can make use of the fact that a line can be linearly ordered or "directed" in two possible ways. If $<$ is one of these two orderings, then we shall say that a point A *precedes* point B if $A < B$. We may also write $B > A$. The details are described in Exercise 20. As usual, $A \leqslant B$ means $A = B$ or $A < B$. We can now rephrase the continuity postulate.

> *Postulate of Continuity (Dedekind's Postulate).* Let ℓ be a line and let $<$ be one of its two possible orderings. If the points of ℓ are divided into two non-empty disjoint sets, so that every point of the first set precedes every point of the second, then there is a unique point C (which may belong to either set) with the property that every point preceding C is in the first set and every point preceded by C is in the second set.

(Hilbert did not adopt Dedekind's Postulate as one of his axioms, but chose instead a pair of assumptions which, taken together, can be shown equivalent to Dedekind's Postulate.)

It is not difficult to see that the analogous verisons of Dedekind's Postulate hold when the word "line" is replaced by "ray" or by "segment." As we shall see shortly, there is a version for angles as well.*

As an application, let us use Dedekind's Postulate (together with some well-known facts of Euclidean geometry) to prove that a segment $\overline{AB}$ joining a point A inside a circle to a point B outside intersects the circle in a unique point.

Let O (assumed given) be the center of the circle and let r be its radius. We divide the segment $\overline{AB}$ into two parts:

$$\mathscr{X} = \{X \text{ on } \overline{AB} \text{ such that } OX < r\}$$
$$\mathscr{Y} = \{Y \text{ on } \overline{AB} \text{ such that } OY \geqslant r\}$$

Since $OA < r < OB$, $\mathscr{X}$ and $\mathscr{Y}$ are non-empty. Moreover, if Z is a point on $\overline{AB}$ which precedes (in the ordering $A < B$) a point X of $\mathscr{X}$, then

---

*Modern definitions of such terms as segment, ray, angle, etc., and their notation in this book, are given in Appendix B.

EUCLID AND THE FOUNDATIONS OF GEOMETRY

since OA < r and OX < r, it follows that OZ < r (Exercise 1), and so Z is in $\mathscr{X}$. Therefore, every point of $\mathscr{X}$ precedes every point of $\mathscr{Y}$.

Applying Dedekind's Postulate to the segment $\overline{AB}$, we obtain a point C separating $\overline{AB}$ into $\mathscr{X}$ and $\mathscr{Y}$. It remains to show that C lies on the circle, i.e., OC = r.

If OC were less than r, we would choose a point S in segment $\overline{CB}$ such that CS < r-OC (Fig. III-8). But then OS < OC + CS < r. This is impossible, since S is in $\mathscr{Y}$. Similarly, it can be proved that OC > r leads to a contradiction as well. Hence OC = r. The uniqueness of C is easy to see.

A consequence of Dedekind's Postulate is the Axiom of Archimedes (a theorem rather than an axiom in our treatment), which, stated informally, says that given two segments, some finite multiple of the first is greater than the second; e.g., by joining together sufficiently many meter sticks, we may eventually reach a distant galaxy (cf. Section II-3).

Theorem III-13 (Axiom of Archimedes)

Let ℓ be a line and let < be one of its two possible orderings. On ℓ, let A, A_1, A_2, A_3, , , , be a sequence of points such that

(i) $A < A_1 < A_2 < A_3 \cdots$
(ii) $AA_1 = A_1 A_2 = A_2 A_3 = \cdots$

Then for any B > A, there exists n such that $A_n > B$.

Proof: Suppose, on the contrary, that $A_n < B$ for every n (if $A_n = B$ for some n, then $A_{n+1} > B$, and the conclusion of the

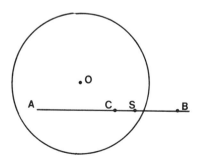

Figure III - 8.

theorem holds). We partition ℓ into two parts 𝒳 and 𝒴 as follows. Let 𝒴 = {Y on ℓ such that A_n < Y for every n}. 𝒳 consists of all other points of ℓ, i.e., 𝒳 = {X on ℓ such that X ⩽ A_n for *some* n}. Then 𝒳 contains every A_n, and 𝒴 contains B, by our assumption. Since every point of 𝒳 precedes every point of 𝒴, there is a unique point C which separates 𝒳 from 𝒴.

A_n < C for every n, since otherwise some A_{n+1} would be preceded by C; an impossibility (see Fig. III-9). Let D be the unique point satisfying D < C and DC = AA_1. Since no A_n can equal D (why?), we have D < A_m < C, for some m (else D in 𝒴). Further, since A_{m+2} < C, we have D < A_m < A_{m+2} < C, for some m. Hence the segment $\overline{A_m A_{m+2}}$ is entirely contained in the segment $\overline{DC}$. This is impossible because the former is twice the size of the latter.

Accordingly, our initial supposition is untenable, and B < A_n for some n.

As is the case with points on a segment, the rays interior to an angle α = ∡AOB may be linearly ordered in two possible ways. For example, if $\overrightarrow{OP}$ and $\overrightarrow{OQ}$ are two interior rays, we can say that $\overrightarrow{OP}$ *precedes* $\overrightarrow{OQ}$, or that $\overrightarrow{OP}$ < $\overrightarrow{OQ}$, if ∡AOP < ∡AOQ. There is a one-to-one order-preserving correspondence between rays interior to α and points of segment $\overline{AB}$: each ray is associated with its intersection with segment $\overline{AB}$. Since any partition of the set of interior rays corresponds to a partition of segment $\overline{AB}$, and vice verse, the following is readily apparent.

Theorem III-15 (Dedekind's Postulate for Angles)

Let α be an angle less than a straight angle, and let < be one of the two possible orderings of its interior rays. If these rays are divided into two non-empty disjoint sets such that every ray of the first set precedes every ray of the second set, then there exists a unique ray which is preceded by every ray of the first set and which precedes every ray of the second set.

Figure III - 9.

EUCLID AND THE FOUNDATIONS OF GEOMETRY

A similar version of the theorem holds for a circular arc, whose points are in obvious one-to-one correspondence with rays emanating from the center of the circle.

C. Infinitude of the Line

Euclid's second postulate asserts that it is possible "to produce [i.e., to extend] a finite straight line continuously [i.e., indefinitely] in a straight line." If we take this literally, a line is certainly without end or "boundary-less," but not necessarily infinite in length. For example, the "lines" on a sphere are its great circles, which, though finite in length, are without end. The distinction between boundary-less and infinite space was pointed out by Riemann in 1854 during his inaugural address on the occasion of his appointment to the faculty of Göttingen. The title of this lecture was "On the Hypotheses Which Lie at the Foundation of Geometry," and the pertinent section might be paraphrased as follows.

> In the application of geometry to space in the large, we must distinguish between unboundedness in extent and infinitude in measure. That space is without boundary and three-dimensional is an empirical certainty that is continually reaffirming itself. (No matter how far we travel, we never encounter a boundary marking the end of space.) This, however, by no means implies that space is infinite (in total volume). From the assumed possibility of motion of bodies without distortion, (i.e., rigid motions) it follows only that space has constant curvature. However, this curvature could conceivably have a positive value, and though this value might be exceedingly small, space would nevertheless be curved and closed in on itself—like a three-dimensional analog of the surface of a sphere—and therefore finite.

The meaning of these remarks will be made clearer in Chapter VIII.

One crucial place where Euclid tacitly assumes that lines are infinite is his proof of Proposition I,16, the *Exterior Angle Theorem*:

> In any triangle, if one of the sides be produced, then the exterior angle is greater than either of the interior and opposite angles.

To prove $\angle ACD > \angle BAC$ in Figure III-10, let E be the midpoint of $\overline{AC}$. Extend $\overline{BE}$ through E to F so that EF = BE. Then $\triangle ABE \cong \triangle CFE$, so that $\angle BAC \cong \angle ACF$. But $\angle ACD > \angle ACF$. Therefore, $\angle ACD > \angle BAC$. Similarly, by extending AC through C to a point G, we can show that $\angle BCG \, (\cong \angle ACD) > \angle ABC$.

The proof seems to depend only on Euclid's first and second postulates, the existence of a segment's midpoint, and the SAS theorem (as well as the Common Notions). All of these assumptions

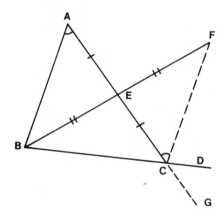

Figure III - 10.

hold on a sphere (where "lines" are great circles). Yet, as the figures below illustrate, the proof breaks down on the sphere. In fact, the exterior angle theorem is false in spherical geometry! Therefore, Euclid's proof must depend on some additional, unstated assumption.

Perhaps the hidden assumption is that two lines cannot meet in more than one point. This would rule out the sphere and prevent $\overleftrightarrow{BF}$ from meeting or crossing $\overleftrightarrow{BC}$ in a point other than B (as in Fig. III-11). Euclid's first postulate states that a line may be drawn from a point to any other point. He does not state that the line is unique, but it is generally agreed that is what he intended. This may be

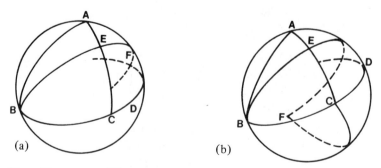

Figure III - 11. (a) If $\overline{BF}$ = a semicircle, F lies on "line" CD and ∡ACD ≅ ∡ACF ≅ ∡BAC. (b) If BF > a semicircle, then ∡ACD < ∡ACF ≅ ∡BAC.

inferred from his assertion in the proof of I,4 that two lines cannot enclose a space. Therefore, if we insert in Euclid's P1 the word "unique" before the word "line," then perhaps that is all that is needed to patch up Euclid's proof.

Unfortunately, this is not the case, as we can see from the *projective plane,* $\mathbb{P}^2$, obtained from the sphere by identifying antipodal points. Specifically, a "point" of $\mathbb{P}^2$ is defined to be a pair of antipodal points on the sphere (i.e., a pair of points that are opposite ends of a diameter), such as the pair formed by the north and south poles. A "line" in $\mathbb{P}^2$ is the set consisting of all antipodal point pairs that lie on the same great circle. Two such pairs determine a unique plane through the sphere's center and so a unique "line." Therefore, Euclid's P1, with the uniqueness requirement, holds in $\mathbb{P}^2$. Moreover, P2 holds also: lines are endless (though finite in length), as on the sphere.

Nevertheless, when $\overline{BE}$ in Figure III-11 is a half-semicircle on the sphere, then B and F will be identified (determine the same point) on $\mathbb{P}^2$, and so Euclid's proof fails here also, even though two lines cannot meet in more than one point. (For $\overline{BE}$ longer, F could even end up on $\overline{BE}$!) Consequently, Euclid's proof of I,16 would seem to require for its validity the assumption of *infinitude of the line,* and this property is indeed implicit in any of the modern postulational foundations for Euclidean geometry.

D. Pasch's Axiom

There is another difficulty in the proof of I,16. How does Euclid know that F is in the interior of ⊰ACD in Figure III-8? Even if lines are assumed infinite and never close up upon themselves, it still must be proved that F is within ⊰ACD. Such concepts as *interior, exterior,* and *between* are not spelled out by Euclid. Whenever he needs to know that a ray passes through the interior of a figure, or that a certain point is between two others, he relies on the visual evidence of his diagrams. But diagrams can mislead (see Exercise 5), and still another assumption is needed to shore up many of Euclid's arguments.

One form of this assumption is the Axiom of Pasch (1882), which Hilbert includes in his postulates. This Axiom guarantees that if a line enters a triangle through a side, then it must exit through one of the other sides.

Pasch's Axiom. Let A, B, and C be three distinct points not all lying on the same line, and let ℓ be a line not containing A, B, or C. Then if ℓ passes through a point of segment $\overline{AB}$, it will also pass through a point of segment $\overline{BC}$ or a point of segment $\overline{AC}$.

(Note: When Pasch's Axiom is being discussed in the context of solid geometry, it must be stipulated that ℓ lies in the plane of A, B, and C. In the present context of plane geometry, this assumption is omitted. In this chapter, "plane" is not one of our undefined terms, although some authors use the words "the plane" to denote the collection of all the points.)

It is not hard to show that ℓ can intersect only one of the segments $\overline{BC}$, $\overline{AC}$ (Exercise 7). Pasch's Axiom enables us to define the two half-planes determined by a line, as well as the interior of an angle or of a triangle. This can best be seen if we show (below) that Pasch's Axiom is equivalent (in the presence of the other postulates of Hilbert) to the Separation Axiom.

Separation Axiom. A line ℓ separates the points which are not on ℓ into two classes such that

   a. if A and B are two points of the same class, then segment $\overline{AB}$ does not intersect ℓ;
   b. if A and B are two points of different classes, then segment $\overline{AB}$ does intersect ℓ.

The two classes determined by ℓ are called *half-planes*. If X is in one of these half-planes, we may call this half-plane the X half-plane of ℓ. The interior of an angle, ∡BAC ($<\pi$), is defined to be the collection of all points common to the B half-plane of $\overleftrightarrow{AC}$ and the C half-plane of $\overleftrightarrow{AB}$. The interior of a triangle, △ABC, is the collection of all points common to the interiors of any two (and hence all three) of its interior angles.

Theorem III-16

On the assumption of Hilbert's postulates I, 1, 2, and 3, and II, 1, 2, and 3, Pasch's Axiom is equivalent to the Separation Axiom (each implies the other).

Proof: Assume Pasch's Axiom. Let X be any point not on ℓ. Let

   $\mathscr{X}$ = {A such that A is not on ℓ and ℓ does not meet segment $\overline{AX}$}

EUCLID AND THE FOUNDATIONS OF GEOMETRY

$\mathcal{Y} = \{$B such that B is not on ℓ and ℓ meets segment $\overline{BX}\}$

To see that $\mathcal{X}$ is non-empty, join X to any point Y on ℓ. Then any point of segment $\overline{XY}$ is in $\mathcal{X}$. That $\mathcal{Y}$ is non-empty may be seen by choosing a point on the extension of segment $\overline{XY}$ through Y. By mutual agreement, we shall include X in $\mathcal{X}$.

It is now easy to show, on the basis of Pasch's Axiom that $\mathcal{X}$ and $\mathcal{Y}$ are the two classes described in the Separation Axiom. For example, suppose A and A' belong to $\mathcal{X}$. Then ℓ meets neither of the segments $\overline{AX}$ and $\overline{A'X}$. But then ℓ cannot meet segment $\overline{AA'}$; otherwise a contradiction of Pasch's Axiom would result (the case where A, A', and X are collinear is easily handled). Similarly, by Exercise 7, if B and B' belong to $\mathcal{Y}$, then ℓ cannot meet segment $\overline{BB'}$. Finally, if A belongs to $\mathcal{X}$ and B to $\mathcal{Y}$, Pasch's Axiom requires that ℓ meet segment $\overline{AB}$. It is not hard to show that the partition of the points of α not on ℓ into two half-planes does not depend on the choice of X (see Exercise 18).

The deduction of Pasch's Axiom from the Separation Axiom is easier and is left to you (Exercise 17).

The following consequence of Pasch's Axiom says that a line entering a triangle through one of its vertices must exit through the opposite side.

Theorem III-17

Let A, B, and C be three distinct points not lying on the same line, and let ℓ be a line which passes through A and contains an interior point of $\triangle ABC$. Then ℓ intersects segment $\overline{BC}$.

Proof: Take any point D on ℓ in the interior of $\triangle ABC$, and let E be any point on segment $\overline{AC}$. Draw $\overleftrightarrow{ED}$. Either $\overleftrightarrow{ED}$ passes through B or not. If not, then by Pasch's Axiom, $\overleftrightarrow{ED}$ meets segment $\overline{AB}$ or segment $\overline{BC}$. If $\overleftrightarrow{ED}$ meets segment $\overline{AB}$ in a point F (Fig. III-12), draw $\overleftrightarrow{FC}$ and apply Pasch's Axiom first to $\triangle EFC$ and then to $\triangle BFC$. If $\overleftrightarrow{ED}$ meets segment $\overline{BC}$ in a point G (Fig. III-13), apply Pasch's Axiom to $\triangle EGC$. If $\overleftrightarrow{ED}$ passes through B, use $\triangle BEC$.

In order to plug the gap in Euclid's proof of the exterior angle theorem (Euclid I,16), we need only prove that ray $\overrightarrow{CF}$ is interior to $\angle ACD$, so that $\angle ACF < \angle ACD$ (Fig. III-14). This follows from the Separation Axiom: F is in the D half-plane of line $\overleftrightarrow{AC}$ (since segments $\overline{BF}$ and $\overline{BD}$ meet line $\overleftrightarrow{AC}$), and F is also in the A half-plane of line $\overleftrightarrow{CD}$ (check this).

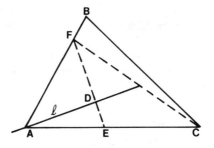

Figure III - 12.

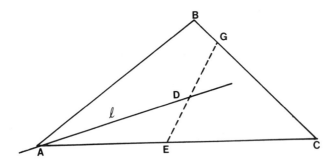

Figure III - 13.

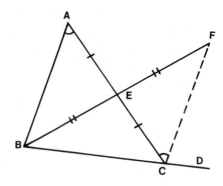

Figure III-14.

Notice that from the inequality $\pi - \angle BCA = \angle ACD > \angle BAC$, it follows that $\angle BAC + \angle BCA < \pi$. Accordingly, in any triangle the sum of any two interior angles is less than a straight angle. This is Euclid's Proposition I,17.

We can now see that the Separation Axiom plays a crucial role in the proof of the Exterior Angle Theorem, as well as in many other Euclidean propositions. We also saw the importance of the assumption that two lines intersect in at most one point. In the two geometries we have seen where lines are finite, these properties do not both hold.

In Chapter IV, we shall examine the controversy surrounding Euclid's fifth postulate, and trace some of the steps that eventually led to the discovery of alternative geometries.

Exercises III-3

In the following problems, you may assume any of the first twenty-eight Propositions of Euclid's Book I, listed in Appendix A.

1. In $\triangle ABC$, suppose $AB < r$ and $AC < r$, for some positive number r. Show that for any z on $\overline{BC}$, $AZ < r$.

2. (a) Suppose $\angle POQ$ is acute, and let A be any point on ray $\overrightarrow{OQ}$. Show that the foot of the perpendicular dropped from A to line $\overleftrightarrow{OP}$ lies on the P side of O (i.e., on ray $\overrightarrow{OP}$.) (b) Prove: if $\angle A$ is the (or a) largest angle in $\triangle ABC$, then the altitude drawn from A meets line $\overleftrightarrow{BC}$ in a point between B and C.

3. In the demonstration that a segment joining an interior point of a circle to an exterior point intersects the circle, it is stated that "$OC > r$ (Fig. III-10) leads to a contradiction as well." Prove this.

4. (The *Hypotenuse-Leg Theorem* or HL). In triangles $\triangle ABC$ and $\triangle DEF$, suppose $\angle B$ and $\angle E$ are right, $AB = DE$, and $AC = DF$. Prove $\triangle ABC \cong \triangle DEF$. (Do not assume the Pythagorean Theorem.)

5. What is wrong with the following "proof" that every triangle is equilateral?

In $\triangle ABC$, let the bisector of $\angle A$ meet the perpendicular bisector of $\overline{BC}$ at E (see the figure at the top of page 120). Draw $\overline{BE}$ and $\overline{CE}$, and drop the perpendiculars $\overline{EF}$ and $\overline{EG}$ from E to $\overline{AB}$ and $\overline{AC}$, respectively. Then $\triangle AEF \cong \triangle AEG$ (by SAA), and so $AF = AG$ and $EF = EG$. Also, $\triangle BDE \cong \triangle CDE$ (by SAS), and so $BE = CE$.

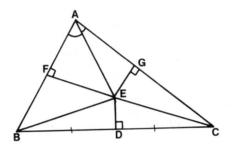

In right triangles △BFE and △CGE, we now have BE = CE and EF = EG. Therefore △BFE ≅ △CGE (by Exercise 4), and so BF = CG. But AF = AG and BF = CG together imply AB = AC. A similar argument (beginning with the bisector of ∡B) can be used to show AB = BC. Hence △ABC is equilateral.

6. Consider the following interpretations for Hilbert's undefined terms.

"Point" = a points (x,y), with $y > 0$, in the Cartesian plane.

"Line" = either an open semicircle with center on the x axis, or an open ray perpendicular to the x axis. [See the figure at the bottom of this page. "Open" means "without end point(s)."]

"Lies on" = "belongs to" in the set-theoretic sense.

"Between": If A, B, and C are "points" of a "line," say "B is between A and C" if, as the "line" is traversed in some direction, the "points" are encountered in the order A, B, C or C, B, A.

"Congruent for segments" means "have equal arc lengths."

"Congruent for angles" means congruent in the ordinary Euclidean sense. (The angle between curves is the angle between their tangents at the point of intersection.)

Which of Hilbert's postulates hold in the above interpretations? For those that do not, give a reason or counterexample. (Remember, points on the x axis are not "points.")

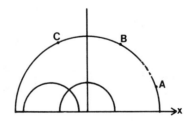

EUCLID AND THE FOUNDATIONS OF GEOMETRY

In Exercises 7 to 16, you may use any of Hilbert's postulates of incidence and order or their consequences, including the Separation Axiom. See Appendix B.

7. Show that a line which does not pass through a vertex of a triangle cannot intersect all three of the triangle's sides. (Hint: suppose a line ℓ intersects the three sides of $\triangle ABC$ in the points D on $\overline{AB}$, E on $\overline{AC}$, and F on $\overline{BC}$. One of these points, say E, must be between the other two. Now find a triangle and line which contradict Pasch's Axiom.)

8. Extend side $\overline{BC}$ of $\triangle ABC$ through C to any point D, and let E be a point between A and C. Prove $\overleftrightarrow{DE}$ meets $\overline{AB}$.

9. Prove that if C ($\neq$ A) is a point of ray $\overrightarrow{AB}$, then the rays $\overrightarrow{AC}$ and $\overrightarrow{AB}$ are one and the same. ("Ray" is defined in Appendix B.)

10. Show that if two collinear rays $\overrightarrow{AB}$ and $\overrightarrow{CD}$ have a point in common other than A or C, then their intersection is a segment or ray.

11. In Figure III-14, prove that F is in the A half-plane of line $\overleftrightarrow{CD}$.

12. Show that if a line through the vertex of a pair of vertical angles contains an interior point of one of them, then it contains an interior point of the other.

13. Prove that if A and B are interior points of an angle, then the entire segment $\overline{AB}$ is contained in the interior of the angle.

14. Prove: if A is an interior point of $\angle POQ$, then every point of ray OA (other than O itself) is an interior point of $\angle POQ$.

15. Prove: if A and B are in the same half-plane of a given line, then at least one of the rays $\overrightarrow{AB}$, $\overrightarrow{BA}$ lies in this half-plane.

16. Fill in the details in this alternate proof of Theorem III-17 (figure on this page).

Extend $\overline{BA}$ through A to any point E and draw $\overline{CE}$. Then ℓ meets side $\overline{BE}$ of $\triangle BCE$ but does not pass through a vertex of this triangle (why?). By Pasch's Axiom, ℓ must meet $\overline{BC}$ or $\overline{CE}$. Since it cannot meet $\overline{CE}$ (why?), it must meet $\overline{BC}$.

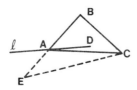

17. Assuming Hilbert's postulates I, 1, 2, 3 and II, 1, 2, 3, and their consequences, deduce Pasch's Axiom from the Separation Axiom. (Note: to prove Pasch's Axiom, you begin by taking its hypothesis as given, i.e., assume a triangle, $\triangle ABC$, and a line ℓ passing through the interior of $\overline{AB}$.)

18. In the proof that Pasch's Axiom implies the Separation Axiom, the two classes $\mathscr{X}$ and $\mathscr{Y}$ were defined with reference to a particular choice of point X. Suppose that X' is a different point not on ℓ and define

$\mathscr{X}' = \{A$ such that A is not on ℓ and ℓ does not meet $\overline{AX'}\}$
$\mathscr{Y}' = \{B$ such that B is not on ℓ and ℓ meets $\overline{BX'}\}$

Show that either $\mathscr{X}' = \mathscr{X}$ and $\mathscr{Y}' = \mathscr{Y}$, or else $\mathscr{X}' = \mathscr{Y}$ and $\mathscr{Y}' = \mathscr{X}$.

19. Show (on the assumption of Hilbert's postulates I, 1, 2, 3 and II, 1, 2, 3) that Theorem III-17 implies Pasch's Axiom.

20. Say that two rays of a given line are *codirectional* if their intersection is a ray. (a) Show that codirectionality is an equivalence relation (cf. Example vi in Section 2). Equivalence classes will be called *directions*. (b) Show that a line has exactly two directions. If $\overrightarrow{AB}$ represents a direction on a line ℓ, and X and Y are distinct points of ℓ, we shall say $X < Y$ if $\overrightarrow{AB}$ and $\overrightarrow{XY}$ are codirectional. (c) Show that the definition of $X < Y$ does not depend on the choice of representative for the direction $\overrightarrow{AB}$. (d) Show that $<$ is transitive and that $X < Y$ excludes $Y < X$.

21. (*The Nested Interval Property*) Prove the following.
Let $\overline{A_1 B_1}$, $\overline{A_2 B_2}$, $\overline{A_3 B_3}$, ... be a sequence of segments on a given line, and suppose (i) $\overline{A_n B_n}$ contains $\overline{A_{n+1} B_{n+1}}$ for every positive integer n; (ii) there does not exist a segment that is contained in every $\overline{A_n B_n}$. Then there exists a unique point C belonging to $\overline{A_n B_n}$ for every n. (Hint: choose an ordering $<$ for the line, and assume, by relabeling if necessary, $A_n < B_n$ for all n. Show $A_m < B_n$ for all m and n. Let $\mathscr{X} = \{X$ such that $X < B_n$ for all $n\}$, $\mathscr{Y} = \{Y$ such that $Y \geq B_n$ for some $n\}$.)

22. (a) Using Hilbert's postulates I, 1-8 and II, 1-4, prove the following spatial analog of Pasch's Axiom. Let A, B, and C be three noncollinear points and let α be a plane which does not contain A, B, or C. Then if α passes through a point of segment $\overline{AB}$, it must also pass through a point of $\overline{AC}$ or a point of $\overline{BC}$. (b) Having proved the

EUCLID AND THE FOUNDATIONS OF GEOMETRY

above, state and prove a spatial separation property analogous to the (plane) separation axiom given in this section. Define *half-space*. [Note: the statement proved in (a) may be viewed as expressing the "3-dimensionality" of space, just as Pasch's Axiom, in the form given in this section, may be viewed as expressing the "2-dimensionality" of the plane.]

23. (For Calculus students only) Show that the following four statements are equivalent. (a) For any positive numbers a and b, there exists an integer N such that $Na > b$. (This is the analytic version of the Axiom of Archimedes.) (b) $\lim_{n \to \infty} \frac{1}{n} = 0$. (c) The positive integers are unbounded, i.e., given any positive number M, there exists an integer n such that $n > M$. (d) $\lim_{n \to \infty} \frac{1}{2^n} = 0$ (essentially the analytic version of the Principle of Exhaustion; cf. Section II-3).

24. Show that Dedekind's Postulate does not hold in the rational plane [model (iv) in Section 1].

IV

HISTORY OF THE PARALLEL POSTULATE

> *... out of nothing I have created a strange new universe.*
> —J. Bolyai

> *What Vesalius was to Galen, what Copernicus was to Ptolemy, that was Lobachevski to Euclid.*
> —William Clifford

1. THE PARALLEL POSTULATE

In standard logic, a *statement* is an assertion that is either true or false, but not both. Accordingly, one way of proving a statement is true is to demonstrate that its falsity is impossible. This is usually accomplished by showing that its falsity entails some contradiction. Here is an example.

Proposition

Let n be an integer. If n^2 is odd, then n is odd.

Proof: Let us suppose, on the contrary, that n were not odd, i.e. even, and try to deduce from this a contradiction. If n were even, we could write n as $n = 2m$, for some integer m. However, then $n^2 = (2m)^2 = 4m^2$, being a multiple of 4, would also be even. This contradicts the hypothesis "n^2 is odd." Hence the supposition that led to this contradiction must be false: accordingly, n is odd.

Notice that we proved "n^2 odd implies n odd" by showing "n *not* odd implies n^2 *not* odd." The above is an illustration of *proof by contradiction,* or the indirect method of proof, wherein we show that denial of the conclusion contradicts the hypothesis. This method depends on the following principle of logic: if P and Q are any statements, then "P implies Q" is a statement logically equivalent to the statement "not Q implies not P" (see Exercise 6).

Expressed with the standard symbols $\sim$ for "not" and $\rightarrow$ for "implies," this says that each of the statements (P $\rightarrow$ Q) and ($\sim$Q $\rightarrow$ $\sim$P) is a logical consequence of the other. If we can prove one of them, then the other holds automatically. $\sim$Q $\rightarrow$ $\sim$P is called the *contrapositive* of P $\rightarrow$ Q. (P $\rightarrow$ Q may also be read "if P then Q.")

On the other hand, the *converse* of P $\rightarrow$ Q is the statement Q $\rightarrow$ P, and this is not equivalent to P $\rightarrow$ Q. For example, the statement "if n is a prime positive integer, then n is at least two" is true, but its converse is obviously false: if n is at least two, it need not be prime. See Table IV-1.

In what follows, it will be important to bear in mind the distinction between converse and contrapositive. We shall focus our attention on the Euclidean parallel postulate. The converse of this postulate is deducible from the other postulates of Hilbert, but this by no means guarantees the provability of the postulate itself.

The definitions, postulates, and propositions of Book I of the *Elements* are given in Appendix A. Euclid gave the following definition of parallelism.

Parallel straight lines are straight lines which, being in the same plane and being produced indefinitely in both directions, do not meet one another in either direction. (Euclid's "straight line" corresponds to our undefined term "line.")

Euclid's fifth postulate (P5), the famous Parallel Postulate, is the following statement.

Table IV - 1

Conditional Statement	P $\rightarrow$ Q	
Its Contrapositive	$\sim$Q $\rightarrow$ $\sim$P	logically equivalent
Its Converse	Q $\rightarrow$ P	
Its Inverse	$\sim$P $\rightarrow$ $\sim$Q	logically equivalent

THE PARALLEL POSTULATE

> If a straight line falling on two straight lines make the interior angles on the same side less than two right angles, the two straight lines, if produced indefinitely, meet on that side on which are the angles less than the two right angles.

This rather complicated postulate stipulates that if $\alpha + \beta < \pi$ in Figure IV-1, then lines m and n meet if extended sufficiently far to the right. (Throughout this text the symbol π is used to denote a straight angle, i.e., the figure formed by two opposite rays with common initial point.) Line ℓ is called a *transversal*.

This postulate is far more complex than Euclid's other postulates, and is certainly not "self-evident." A first grader could probably understand the first four postulates, but not this one. Proclus (5th c. A.D.), who wrote a *Commentary* on Euclid, described P5 as "alien to the special character of postulates" and an assertion that properly should be a theorem derived from the remaining postulates. However, as we shall see, all efforts over two thousand years to deduce the parallel postulate from the remaining postulates of Euclidean geometry met with failure.

Labeling the hypothesis and conclusion P and Q, we may write P5 symbolically as P → Q. The converse, Q → P, says that if the lines meet, then $\alpha + \beta < \pi$. This is essentially the content of Euclid's Prop. I, 17, which is an immediate consequence of the Exterior Angle Theorem (Euclid I, 16) and does not depend on P5 for its proof (cf. Section III-3).

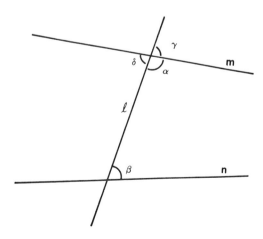

Figure IV - 1.

Euclid I, 17. In any triangle, the sum of any two angles is less than two right angles.

(Q → P)

If the interior angles are together less than a straight angle on *neither* side of a transversal, then on each side of the transversal the interior angles are supplementary. Therefore the following proposition is essentially ~P → ~Q, which is the contrapositive of (and hence equivalent to) the *converse* of P5. (~P → ~Q is also called the *inverse* of P → Q.)

Euclid I, 28. If a straight line falling on two straight lines makes the exterior angle equal to the interior and opposite angle on the same side ($\beta \cong \gamma$ in Figure IV-2), or the interior angles on the same side equal to two right angles, then the straight lines will be parallel to one another.

(~P → ~Q)

None of the first twenty-eight propositions in Book I of the *Elements* depend for their proofs upon the parallel postulate. Proposition I, 29 is the first to require it.

Euclid I, 29. A straight line falling on parallel straight lines makes . . . the interior angles on the same side (together) equal to two right angles.

(~Q → ~P)

This is simply the contrapositive of P5, and so an immediate consequence of it. The proof runs as follows: If the conclusion of I, 29 were false, then on one of the two sides of the transversal the sum of the interior angles would be less than two right angles. By P5, the two lines would then necessarily meet, contradicting the hypothesis (Table IV-2).

A common and familiar substitute for P5 is generally referred to as Playfair's Axiom (1795, John Playfair, Scottish physicist and mathematician). The axiom appeared in equivalent form in Proclus' commentaries and is certainly implicit in Euclid.

THE PARALLEL POSTULATE

Table IV - 2

Euclid's P5	P → Q	α + β < π → m,n meet	logically equivalent
Euclid's I,29	~Q → ~P	m ∥ n → α + β = π	
Euclid's I,17	Q → P	m,n meet → α + β < π	logically equivalent
Euclid's I,28	~P → ~Q	α + β = π → m ∥ n	

Playfair's Axiom

Through a given point not on a given line there can be drawn (in the plane of this point and line*) only one line parallel to the given line.

Since two lines with a common perpendicular are parallel (Euclid I, 28), we know (in the absence of *any* parallel postulate) that there is at least one parallel to a given line through a given external point. The equivalence of P5 and Playfair's Axiom (on the assumption of the remaining postulates of geometry) can then be proved (see Exercise 1 of Section IV-2). Hilbert adopts Playfair's form of the parallel postulate, rather than Euclid's, as do most modern geometry texts.

Exercises IV-1

1. Write the converse and the contrapositive of each of the following statements.

   a. If it rained this morning, then the streets are wet.
   b. If a triangle is equilateral, then all its angles are congruent.
   c. If $x < y$, then $x \neq y$.
   d. If two angles are vertical angles, then they are congruent.

2. (a) What is the contrapositive of ~Q → P? (b) What is the contrapositive of the contrapositive of P → Q?

3. Show that "if a straight line falling on two straight lines makes the interior angles on the same side less than two right angles, then the two straight lines, if produced indefinitely," cannot possibly meet on the side of the transversal opposite the two angles. (Thus the last 14 words of Euclid's P5 are superfluous.)

---

*In the present context of plane geometry, this phrase in parentheses is unnecessary.

4. Using any of Euclid's first 16 propositions, prove (a) if, in $\triangle ABC$, $AC > BC$ then $\angle B > \angle A$ (Euclid I, 18); (b) if, in $\triangle ABC$, $\angle B > \angle A$ then $AC > BC$ (Euclid I, 19); (c) in any $\triangle ABC$, $AB + AC > BC$ (Euclid I, 20). (Hint: extend $\overline{BA}$ through A to D so that $AD = AC$.)

5. Prove that if two lines are cut by a transversal so that alternate interior angles are congruent (e.g., $\beta \cong \delta$ in Fig. IV-1), then the two lines have a common perpendicular.

6. (For students with some previous exposure to symbolic logic.) The truth values of a conditional statement $P \rightarrow Q$ (where P and Q are any statements) are given in the third column of the table below: $P \rightarrow Q$ is false only when P

P	Q	$P \rightarrow Q$	$\sim Q$	$\sim P$	$\sim Q \rightarrow \sim P$	$\sim P \vee Q$
T	T	T				
T	F	F				
F	T	T				
F	F	T				

is true and Q is false. Complete the remaining columns in the table and verify that the statements $\sim Q \rightarrow \sim P$ and $\sim P \vee Q$ (not P or Q) are logically equivalent to (i.e., have the same truth tables as) $P \rightarrow Q$.

7. Write the truth tables for $Q \rightarrow \sim P$ and $P \rightarrow \sim Q$. What do you notice? Explain.

2. ABSOLUTE GEOMETRY

Since, as we saw in Chapter III, Euclid's postulate set is incomplete, we shall take as a foundation for Euclidean geometry the postulate set of Hilbert.

Many propositions of a quantitative, or "metric" nature depend on the parallel postulate for their proof. Chief among these are the Pythagorean theorem and the theorem stating that the sum of the interior angles of a triangle equals two right angles.

On the other hand, there is a large collection of theorems which are independent of the parallel postulate and can be derived from the rest of Hilbert's postulates. Such theorems are said to be theorems of *absolute geometry*, because, as we shall see, they hold true both in Euclidean and in Lobachevskian geometry.

ABSOLUTE GEOMETRY

Definition IV-1

Absolute geometry is the collection of all statements deducible from Hilbert's postulates with the Axiom of Parallels deleted. It is the body of theorems common to both Euclidean and Lobachevskian geometry.

Belonging to absolute geometry are the first 28 propositions of Euclid (see Appendix A), including the triangle congruence theorems (SAS, ASA, SSS, and SAA). Also belonging to absolute geometry are many propositions of a non-metric nature that express an order relation or symmetry. For example, "if two sides of a triangle are not congruent, the angle opposite the greater side is the greater," "the base angles of an isosceles triangle are congruent," and "the perpendicular bisector of a chord of a circle is a diameter."

As we proceed further, you will gradually develop an intuition concerning which theorems do or do not belong to absolute geometry. Let us now find out how far we can go without a parallel postulate. (The theorems below are results of absolute geometry.)

Recall that in any triangle, the sum of any two angles is less than two right angles (Euclid I, 17). What about the sum of all three?

Theorem IV-2

In any triangle, the sum of the three angles is less than or equal to two right angles.

Proof: Suppose, on the contrary, that there exists a triangle, $\triangle ABC$, whose angle sum, denoted $\mathscr{S}(ABC)$, exceeds π, i.e., $\mathscr{S}(ABC) = \pi + \phi$, where $\phi > 0$. Let α be the (or a) minimal angle in $\triangle ABC$. We may assume (by relabeling if need be) that $\alpha = \measuredangle BAC$. Let D be the midpoint of $\overline{BC}$. Draw $\overline{AD}$ and extend it through D to B_1 so that $AD = DB_1$. Since $\triangle ABD \cong \triangle B_1 CD$, it follows that $\triangle ABC$ and $\triangle AB_1 C$ have the same angle sum ($\alpha_1 + \alpha_2 + \beta + \gamma$) (in Fig. IV-2).

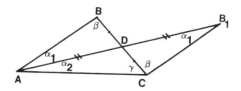

Figure IV - 2.

Moreover, since $\alpha = \alpha_1 + \alpha_2$, at least one of the two angles α_1, α_2 must be $\leq \alpha/2$. Consequently, the minimal angle in $\triangle AB_1C$ is $\leq \alpha/2$. $\triangle AB_1C$ is called the *first derived triangle* (of $\triangle ABC$).

Beginning with any given triangle (such as $\triangle ABC$), the above construction produces a new triangle ($\triangle AB_1C$) with the same angle sum but with minimal angle no more than half the minimal angle in the original triangle. We may reapply this process to $\triangle AB_1C$ and obtain a *second derived triangle* with the same angle sum, but minimal angle $\leq \alpha/4$.

Iterating n times, we obtain a triangle with angle sum $\pi + \phi$ but minimal angle $\leq \alpha/2^n$. If n is large enough, the latter will be smaller than ϕ.* This forces the remaining two angles to have a sum $> \pi$, which contradicts Euclid I, 17. Therefore no triangle can have an angle sum greater than two right angles.

Definition IV-3

The *defect* of a triangle is the difference between two right angles and the angle sum of the triangle. We denote the defect of $\triangle ABC$ by $\mathscr{D}(ABC)$. Thus

$$\mathscr{D}(ABC) = \pi - \mathscr{S}(ABC)$$

In Euclidean geometry, the defect of every triangle is zero (Euclid I, 32). Theorem IV-2 states only that $\mathscr{D} \geq 0$ in every triangle (absolute geometry).

Lemma IV-4

If a line is drawn from vertex A of $\triangle ABC$ to a point D on the opposite side, then $\mathscr{D}(ABC) = \mathscr{D}(ABD) + \mathscr{D}(ACD)$.

Proof: Compute both sides of the equation, using the labeling in Figure IV-3. This lemma is generalized in Exercise 2.

Because defects are additive, it is usually simpler to work with them directly rather than with angle sums.

Theorem IV-5

If the sum of the angles of some triangle equals two right angles, then the same is true of every triangle.

---

*Theorem IV-2 depends critically upon the Axiom of Archimedes (see Exercise 23 of Section III-3).

ABSOLUTE GEOMETRY 133

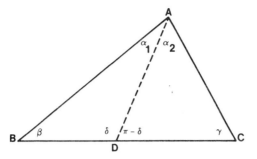

Figure IV - 3.

Proof: Suppose $\mathscr{S}(ABC) = \pi$, or equivalently, $\mathscr{D}(ABC) = 0$. Assume $\angle BAC$ is the (or a) largest angle of $\triangle ABC$. Then, by Exercise 2 of Section III-3, the altitude $\overline{AD}$ (Fig. IV-4) divides $\triangle ABC$ into two right triangles, $\triangle ABD$ and $\triangle ACD$, both with defect zero, by Lemma IV-4. Because $\mathscr{S}(ABD) = \pi$, we may adjoin to $\triangle ABD$ a congruent right triangle in such a way that their hypotenuses coincide and together they form a rectangle R. Because all four angles of R are right and opposite sides are equal, we may arrange rectangles congruent to R in brick wall fashion—as many as we wish—to obtain an arbitrarily large rectangle with angle sum 2π.

This rectangle is divided by a diagonal into two right triangles, each of which must have angle sum π (why?). What we have accomplished, then, is to show that if there is *some* triangle with zero defect, then there are arbitrarily *large* right triangles with zero defect.

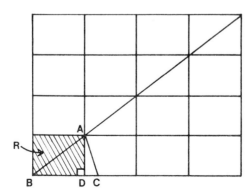

Figure IV - 4.

To show now that *every* triangle has defect zero, it suffices to show this for right triangles, since, as we saw above, every triangle is divided into two right triangles by an altitude.

To this end, let △DEF be an arbitrary right triangle (Fig. IV-5). We know there exists a right triangle, T, with zero defect and with each leg greater than both those of △DEF. Extend $\overline{DE}$ and $\overline{DF}$ to G and H, respectively, so that △DHG ≅ T. Draw EH. Then

$$0 = \mathscr{D}(DGH) = \mathscr{D}(GEH) + \mathscr{D}(DEH)$$
$$= \mathscr{D}(GEH) + \mathscr{D}(FEH) + \mathscr{D}(DEF)$$

Since defects are non-negative in absolute geometry, this implies $\mathscr{D}(DEF) = 0$.

Theorems IV-2 and IV-5, which are often called Legendre's first and second theorems, together yield the following.

Theorem IV-6

In any model for the postulates of absolute geometry, exactly one of the following holds:

a. in every triangle the sum of the angles equals two right angles;
b. in every triangle the sum of the angles is less than two right angles.

(We saw earlier that on a sphere, where "lines" are great circles, and in the projective plane—see Section III-3—the sum of the angles in

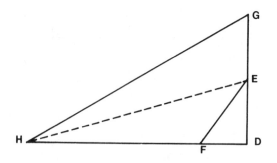

Figure IV - 5.

ABSOLUTE GEOMETRY

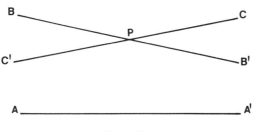

Figure IV - 6.

any triangle exceeds two right angles. However, these geometries are not models for the postulates of absolute geometry, since lines are finite and the order axioms fail.)

Since in absolute geometry there is always at least one parallel to a given line through a given external point, the negative of Playfair's Axiom would postulate the existence of more than one such parallel. This is the assumption which replaces P5 in the geometry discovered by N. Lobachevsky. (Actually, the negation of Playfair's Axiom is the statement that there exist for *some* line ℓ and *some* point P not on ℓ at least two lines through P which do not meet ℓ. However, it will follow from Theorem IV-7 that *some* can be replaced by *any*.)

Lobachevsky's Parallel Postulate

Through a point not lying on a given line, there exist (in the plane of this point and this line*), at least two lines which do not meet the given line.

Suppose, in Figure IV-6, that lines $\overleftrightarrow{BB'}$ and $\overleftrightarrow{CC'}$ do not meet $\overleftrightarrow{AA'}$. (Assume the labeling is such that the perpendicular to $\overleftrightarrow{AA'}$ through P passes through the vertical angles $\angle C'PB'$ and $\angle BPC$.) It is easily seen that any line through P interior to angles $\angle BPC'$ and $\angle B'PC$ will likewise fail to meet $\overleftrightarrow{AA'}$ (Exercise 6). Hence the Lobachevskian postulate actually furnishes an infinite number of lines through P which do not meet $\overleftrightarrow{AA'}$.

Before proceeding, let us look at some models for Lobachevsky's postulate.

---

*In the context of plane geometry, the phrase in parentheses may be omitted.

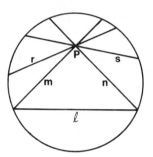

Figure IV - 7.

Example 1. (*The Klein-Beltrami Disk Model*) Let Σ be a fixed circle. Interpret "point" to mean "interior point of Σ." Interpret "line" to mean "open chord of Σ," i.e., a chord with end points deleted. In Figure IV-7, P is a "point" not on "line" ℓ, and m, n, r, and s are "lines" through P which do not "meet" ℓ. (Remember, points on Σ do not count as "points," and so m and ℓ really do not have a "point" in common.)

Example 2. (*The Poincare Disk Model*) Let Σ be a fixed circle. Interpret "point" to mean "interior point of Σ," just as in Example 1. But define "line" to mean "the set of interior points of Σ that lie on a circle which intersects Σ orthogonally." We shall include open diameters—which can be viewed as circular arcs of infinite radius—as "lines" also. See Figure IV-8.

[Henri Poincaré (1854-1912) was a French mathematician and physicist who made numerous contributions to analysis, topology, geometry, and the philosophy of science, as well as to the foundations of relativity theory. See also Section VIII-3.]

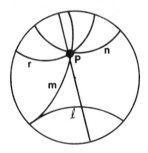

Figure IV - 8.

ABSOLUTE GEOMETRY

Example 3. (*The Poincare Upper Half-Plane Model*) The interpretation of Exercise 6 of Section III-3 is also a model for Lobachevsky's postulate. Two semicircles, or a line and a semicircle, which are tangent to each other at a point of the x axis do not have a "point" in common. Hence no two of the "lines" r, s, t in Figure IV-9 have a common "point."

With the obvious interpretation of "between," all three of these models satisfy Hilbert's postulates of incidence (I, 1, 2, and 3) and order, and Dedekind's Postulate. Of course, if congruence of segments is interpreted to mean equality of arc length, then the congruence postulates obviously fail. However, in each of these interpretations, it is possible to define congruence in another way, so that all of Hilbert's postulates hold except for the Axiom of Parallels, which is replaced by Lobachevsky's postulate. These three examples (with the appropriate definitions of congruence) are therefore models for Lobachevskian geometry. (They were unknown in Lobachevsky's day.) We shall develop these models further in subsequent chapters.

The next result establishes the connection between the angle sum of a triangle, and the choice of parallel postulate.

Theorem IV-7

(a) If the sum of the angles in every triangle is $< \pi$, then Lobachevsky's postulate holds. (b) If the sum of the angles in every triangle $= \pi$, then Euclid's postulate holds.

Proof. Let P be a point not lying on line ℓ, and let Q be the foot of the perpendicular from P to ℓ. Through P draw line $\overleftrightarrow{PA}$ perpendicular to $\overleftrightarrow{PQ}$, as in Figure IV-10. By Euclid I, 28, $\overleftrightarrow{PA}$ does not meet ℓ.

Now suppose that in every triangle the sum of the angles is $< \pi$. Draw $\overleftrightarrow{PR}$ from P to any point R (other than Q) on ℓ.

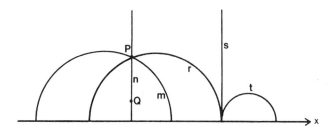

Figure IV - 9.

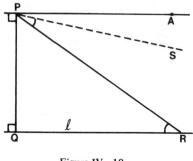

Figure IV - 10

On the side of $\overleftrightarrow{PR}$ opposite Q, draw ray $\overrightarrow{PS}$ so that ∡RPS ≅ ∡PRQ. By Euclid I, 28, $\overleftrightarrow{PS}$ does not meet ℓ. Moreover, since

$$\pi > \mathscr{S}(PQR) = \pi/2 + \text{∡PRQ} + \text{∡RPQ}$$
$$= \pi/2 + \text{∡RPS} + \text{∡RPQ} = \pi/2 + \text{∡SPQ}$$

then ∡SPQ < $\pi/2$, so ray $\overrightarrow{PS}$ is interior to ∡APQ. Hence there are at least two *distinct* lines ($\overleftrightarrow{PA}$ and $\overleftrightarrow{PS}$) through P which do not meet ℓ. This establishes (a).

Next, assume that the sum of the angles in every triangle equals π. On one side of $\overleftrightarrow{PQ}$, mark points $Q_1, Q_2, Q_3, \ldots$ on ℓ such that $QQ_1 = PQ$, $Q_1Q_2 = PQ_1$, $Q_2Q_3 = PQ_2$, etc. (Fig. IV-11). $\triangle PQQ_1$ is an isosceles right triangle with angle sum π. Therefore ∡APQ_1 ≅ ∡$QPQ_1 = \pi/4$.

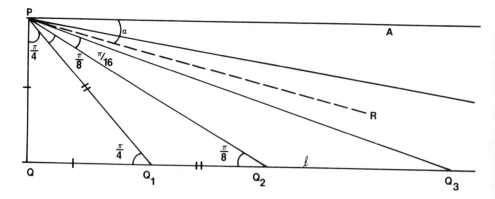

Figure IV - 11.

ABSOLUTE GEOMETRY

Our present assumption is equivalent to the statement that in any triangle an exterior angle equals the sum of the two opposite interior angles. From repeated application of this, we can deduce that

$$\angle APQ_2 = \pi/8, \quad \angle APQ_3 = \pi/16, \quad \angle APQ_4 = \pi/32, \ldots$$

and, in general,

$$\angle APQ_n = \pi/2^{n+1}$$

Consequently, we may construct (on either side of $\overleftrightarrow{PQ}$), lines through P which meet ℓ and which make arbitrarily small angles with line $\overleftrightarrow{PA}$. We now shall show that $\overleftrightarrow{PA}$ is the only line through P which fails to meet ℓ.

Let $\overleftrightarrow{PR}$ be a line through P distinct from $\overleftrightarrow{PA}$. Then on one side of $\overleftrightarrow{PQ}$, $\overleftrightarrow{PR}$ makes an acute angle with $\overrightarrow{PQ}$ and an angle α with ray $\overrightarrow{PA}$. Utilizing the construction above (on the appropriate side of $\overleftrightarrow{PQ}$), with n so large that $\pi/2^{n+1} < \alpha$, we obtain a point Q_n on ℓ such that $\overleftrightarrow{PR}$ passes through the interior of $\angle QPQ_n$ of $\triangle QPQ_n$. By Theorem III-17, $\overleftrightarrow{PR}$ intersects QQ_n. This establishes Playfair's Axiom, which is equivalent to P5. Therefore (b) is proved.

Exercises IV-2

1. Show that on the assumption of absolute geometry, Euclid's P5 and Playfair's Axiom are equivalent.

2. (a) Let D be an interior point of $\triangle ABC$. Show that $\mathscr{D}(ABC) = \mathscr{D}(ABD) + \mathscr{D}(BCD) + \mathscr{D}(ACD)$. (b) Let D, E, and F be interior points of sides $\overline{BC}, \overline{AC},$ and $\overline{AB}$, respectively, of $\triangle ABC$. Prove $\mathscr{D}(ABC) = \mathscr{D}(AEF) + \mathscr{D}(DEF) + \mathscr{D}(BDF) + \mathscr{D}(CDE)$. (More generally, it can be shown that if a triangle is partitioned into any number of subtriangles in any way, then the defect of the whole is the sum of the defect of the parts. See, for example, Borsuk and Szmielew [6, p. 281].)

3. A polygon is called *convex* if the segment joining any two of its interior points lies entirely in the polygon's interior. Prove that in Lobachevskian geometry the sum of the interior angles of a convex polygon of n sides is less than $(n - 2)\pi$. (It is known that this is equally true for non-convex polygons.)

4. How would you define the defect of a polygon of n sides?

5. Prove that in Lobachevskian geometry an angle inscribed in a semicircle is acute.

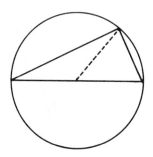

6. Prove: in Figure IV-6 any line passing through the interiors of ⊀BPC′ and ⊀B′PC cannot meet $\overleftrightarrow{AA'}$. (Hint: use Theorem III-17.)

7. (a) In Example 2 of this section, interpret congruence of angles in the "usual" way: the angle between two curves is measured by the angle between their tangent lines at the point of intersection. Sketch an example of a "triangle" in which the angle sum is clearly less than π. (b) Same for Example 3. (Note: for Example 1, however, a different notion of angle congruence is needed to convert that interpretation into a model for Lobachevskian geometry, for in the ordinary interpretation of congruence, the angle sum of any "triangle" is π.)

8. State the contrapositives of Theorem IV-7 a and b. How are these two statements related to the original two? (Note: in absolute geometry, Euclid's and Lobachevsky's postulates are the negations of one another.)

9. In each of the following models, state whether Euclid's postulate holds, Lobachevsky's postulate holds, or neither. In each case, briefly explain why.

(a) "Point" = a point on the surface of an open hemisphere (points on the boundary are excluded). "Line" = an open great semicircle on this hemisphere.

(b) Let P be a fixed point on the surface of a sphere S. "Point" = a point of S distinct from P. "Line" = a circle which lies on S and passes through P, but with P deleted.

(c) Let P be a fixed point of Euclidean 3-dimensional space. "Point" = a straight line through P. "Line" = a plane through P.

(d) "Point" = a line through the origin of $\mathbb{R}^3$ which passes

ABSOLUTE GEOMETRY

through the interior of the cone $z^2 = x^2 + y^2$. "Line" = the set of all "points" which lie on a plane through the origin which passes through the cone's interior.

10. Prove: if there exists a rectangle (quadrilateral with four right angles), then Euclid's postulate holds.

11. Prove: (a) if Euclid's postulate holds, an exterior angle of a triangle equals the sum of the two opposite interior angles; (b) if Lobachevsky's postulate holds, an exterior angle of a triangle is greater than the sum of the opposite interior angles.

12. Assume Lobachevsky's postulate holds. Show that, in the construction of Figure IV-11, $\angle PQ_n Q_{n-1} < \pi/2^{n+1}$ for all $n \geq 2$. (This shows the existence of points S on ℓ such that $\angle PSQ$ is arbitrarily small.)

3. ABSOLUTE LENGTHS

Theorem IV-8

If there exist two similar,* but noncongruent triangles, then Euclid's postulate holds.

Proof. Suppose in triangles $\triangle ABC$ and $\triangle A'B'C'$ that $\angle A \cong \angle A'$, $\angle B \cong \angle B'$, $\angle C \cong \angle C'$, but that $AB > A'B'$. On side $\overline{AB}$ mark point B'' such that $AB'' = A'B'$ (Fig. IV-12). On the C side of $\overleftrightarrow{AB}$ draw ray $\overrightarrow{B''C''}$ so that $\angle AB''C'' \cong \angle B'$. By Euclid I, 28 and Pasch's Axiom, $\overrightarrow{B''C''}$ must meet segment $\overline{AC}$ in a point (which we may as well call)

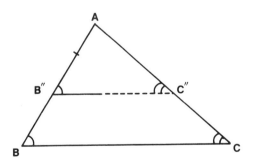 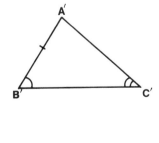

Figure IV - 12.

---

*We call $\triangle ABC$ *similar* to $\triangle A'B'C'$ if $\angle A \cong \angle A'$, $\angle B \cong \angle B'$, and $\angle C \cong \angle C'$.

C''. By ASA, $\triangle AB''C'' \cong \triangle A'B'C'$. Consequently, the angle sum of quadrilateral BCC''B'' is 2π. A diagonal will divide this quadrilateral into two triangles which both have angle sum π. By the previous theorem, Euclid's postulate holds.

Stated in the contrapositive, Theorem IV-8 tells us that in Lobachevsky's geometry a triangle is determined by its three angles: *similar triangles are congruent*! This is a fifth congruence theorem (call it AAA), which holds in Lobachevskian geometry in addition to SAS, ASA, SSS, and SAA, which are absolute theorems.

Let us momentarily focus our attention on equilateral triangles in Lobachevsky's geometry. The preceding gives a one-to-one correspondence between the angles which appear in equilateral triangles and the lengths of their sides. For example, "the side of an equilateral triangle whose angles are all $\pi/4$" is a uniquely determined length. This correspondence is not linear: doubling the sides of a triangle does not result in a doubling of the angles. Nevertheless, it is of deep significance, as we shall see below. The problem of constructing a length corresponding to an angle or the angle corresponding to a length is somewhat complicated, but can be effected solely by means of intersecting lines and circles (i.e., via "straight edge and compass"). These and certain other construction problems of the Lobachevskian plane are solved in Wolfe [53].

In applied Euclidean geometry, we quantify lengths and distances by referring them to standard lengths, such as a meter, a foot, or a mile, etc. Although these units have no special physical significance and are really arbitrarily chosen, they are nevertheless essential for recording and communicating measurements. If all meter sticks, yard sticks, and all other measuring devices, as well as all records of measurements, were suddenly to vanish from the earth, there would be no way for us to reconstruct one meter. Indeed the term "meter" would have no meaning. This is why we have a Bureau of Standards: to maintain the fundamental units in terms of which physical measurements are defined. We summarize all this by saying that "lengths are relative" in Euclidean geometry.

Angles, on the other hand, have a completely different character. There is no need to have a "standard T square" in the Bureau of Standards, since anyone can readily construct a right angle with straight edge and compass (as in Euclid I, 11 and 12), and compare arbitrary angles with submultiples or multiples of a right angle.

ABSOLUTE LENGTHS

Hence we speak of angles as being "absolute," whereas lengths are only "relative." (This use of the word "absolute" has nothing to do with the phrase "absolute geometry.")

Because of Theorem IV-8 the situation in Lobachevsky's geometry is quite different. The correspondence between angles and lengths makes lengths absolute as well as angles. Any length is determined by and constructible from its corresponding angle. For example, in a model for Lobachevskian geometry, there is determined, a priori, a certain length which is the side of an equilateral triangle with all three angles $\pi/4$. This length could be used as a standard for measuring. If three-dimensional space were Lobachevskian, there would exist in space a unique distance—perhaps a few centimeters, perhaps many millions of light years—corresponding to $\pi/4$. (We shall discuss this further in Chapter VIII.)

Exercises IV-3

1. In spherical geometry, are lengths relative or absolute? Explain.

2. There are four basic triangle congruence theorems in absolute geometry: SAS, ASA, SSS, and SAA. (A fifth one, AAA, holds in Lobachevskian geometry, but not in Euclidean.) Why is there no SSA theorem?

4. SACCHERI

The first person known to have attempted a proof of Euclid's P5 (from the other postulates) by the indirect method was the Italian Jesuit priest *Gerolamo Saccheri* (1667-1733), a professor of mathematics at the University of Pavia. By assuming all of Euclid which does not depend on P5 and hypothesizing the falsity of P5, he hoped to reach a contradiction.

In his *Euclides ab omni naevo vindicatus* ("Euclid freed of every flaw"), Saccheri considers a birectangular quadrilateral ABCD, in which $\angle A \cong \angle B = \pi/2$ and AD = BC. ($\overline{AB}$ is called the *base*; $\overline{CD}$ is the *summit*.) We call such a quadrilateral a *Saccheri quadrilateral*. By drawing the diagonals (Fig. IV-13) and using congruent triangles, one easily sees that $\angle C \cong \angle D$. There are three possibilities for these *summit angles*—right, obtuse, or acute—which Saccheri labeled

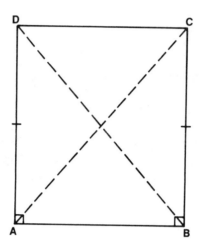

Figure IV - 13.

a. the hypothesis of the right angle,
b. the hypothesis of the obtuse angle,
c. the hypothesis of the acute angle.

(Saccheri proved that if one of these hypotheses holds in a single Saccheri quadrilateral, then it holds in every such quadrilateral.)

The connection between the three hypotheses and the sum of the angles in a triangle is shown in Figure IV-14. Given $\triangle ABC$, let D and E be the midpoints of sides $\overline{AC}$ and $\overline{BC}$, respectively. Drop the perpendiculars $\overline{AF}$, $\overline{BG}$, and $\overline{CH}$ to line $\overleftrightarrow{DE}$. Then by SAA, $\triangle ADF \cong \triangle CDH$, $\triangle BEG \cong \triangle CEH$, and so $AF = CH = BG$. Consequently, ABGF is a Saccheri quadrilateral (with base $\overline{FG}$ and summit $\overline{AB}$) (Fig. IV-14).

Moreover, the sum, $\alpha + \gamma + \beta + \delta$, of the two summit angles (or twice either one) equals the sum of the angles in $\triangle ABC$. Therefore, under the right angle hypothesis, $\mathscr{S}(ABC) = \pi$, and so Euclid's postulate holds. Under the obtuse angle hypothesis, $\mathscr{S}(ABC) > \pi$ (we know this possibility is ruled out in absolute geometry). Under the acute angle hypothesis, $\mathscr{S}(ABC) < \pi$, and so Lobachevsky's postulate holds.

[Figure IV-14 assume $\measuredangle$CDE and $\measuredangle$CED are both acute. However, as you can verify, even when one of these angles is obtuse or right, it is still true that $\mathscr{S}(ABC)$ equals twice the summit angle of ABGF.]

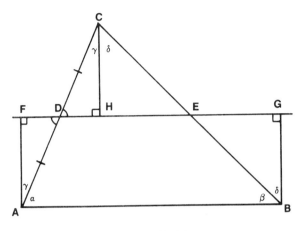

Figure IV - 14.

Saccheri disposed of the hypothesis of the obtuse angle by showing that if it is assumed, one can derive the postulate of Euclid, which he showed was equivalent to the hypothesis of the right angle. His arguments made use of infinitude of the line (in the form of the exterior angle theorem), as well as the Postulate of Archimedes and continuity (cf. Chapter III).

Upon assuming the hypothesis of the acute angle, Saccheri was hard put to find a contradiction. Although he derived some weird consequences of its assumption, such as the existence of a line perpendicular to one side of an acute angle but not meeting the other, the expected contradiction was not at hand. Indeed many of his results became famous theorems of non-Euclidean geometry. He showed, for example, that upon the assumption of the acute angle hypothesis, there would exist, through a given point not on a given line ℓ, two lines which neither meet ℓ nor have a common perpendicular with ℓ, but which approach ℓ asympototically (one in each direction). This anticipated by a century Lobachevsky's theory of parallels (Chapter V).

Relying less on sound logic and more on his faith in the validity of P5, he eventually resorts to vague ideas concerning asymptotic lines with a common perpendicular at infinity to "prove" the "the hypothesis of the acute angle is absolutely false, because it is repugnant to the nature of the straight line." He then gives a second erroneous demonstration based on the idea of equidistant lines. However, the assumption that equidistant straight lines exist is equivalent to Euclid's postulate (cf. Exercise 1).

Saccheri's work went much farther than previous attempts to prove P5. He was the first to catch a glimpse of the "strange new universe" that was to be opened up by the work of Gauss, Bolyai, Lobachevsky, and others. However, as Heath [18, p. 211] notes,

> "Saccheri ... was the victim of the preconceived notion of his time that the sole possible geometry was the Euclidean, and he presents the curious spectacle of a man laboriously erecting a structure upon new foundations for the very purpose of demolishing it afterwards. ..."

A detailed outline of Saccheri's work appears in Bonola [5].

Exercises IV-4

1. Let ℓ and m be two given lines. Show that if the perpendicular distance from a point of ℓ to m is constant, then Euclid's postulate holds.

2. Prove the following theorem of absolute geometry. If, in quadrilateral ABCD, $\angle B \cong \angle C = \pi/2$ and $\angle D > \angle A$, then $AB > CD$. (The side opposite the bigger angle is the bigger.)

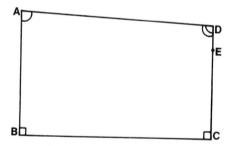

(Hint: Use an indirect proof. Suppose $AB \leq CD$. Mark off $\overline{CE} \cong \overline{BA}$, as in the above figure. Deduce a contradiction.)

3. Show that the line joining the midpoints of the summit and the base of a Saccheri quadrilateral is perpendicular to both. Then, using Exercise 2, prove that in Lobachevskian geometry the summit of a Saccheri quadrilateral is longer than the base.

4. Prove that a perpendicular bisector of a side of a triangle is perpendicular to the line joining the midpoints of the other two sides.

5. Prove that in Lobachevskian geometry, the segment joining the midpoints of two sides of a triangle is shorter than half the triangle's third side.

6. Assume the three perpendicular bisectors of $\triangle ABC$ are concurrent. Show that they are the altitudes of the triangle whose vertices are the midpoints of the sides of $\triangle ABC$.

7. Prove that a line which bisects one side of a triangle and is perpendicular to the perpendicular bisector of another side bisects the third side.

8. Beginning with a Saccheri quadrilateral, construct a triangle one of whose sides is the quadrilateral's summit and whose other two sides are bisected by the quadrilateral's base. Show that all such triangles have the same angle sum.

9. Prove that, in Lobachevskian geometry, two Saccheri quadrilaterals with congruent bases and congruent summit angles are congruent (i.e., the remaining pairs of corresponding parts are congruent).

10. Prove that, in Lobachevskian geometry, two Saccheri quadrilaterals with congruent summits and congruent summit angles are congruent.

5. LAMBERT

In 1763, G. S. Klügel published a dissertation entitled, *Conatuum praecipuorum theoriam parallelarum demonstrandi recensio* ("A review of the main attempts at proving the theory of parallels"), in which he discussed all of the significant attempts to date to prove the Euclidean parallel postulate (including Saccheri's). All were found wanting. Klügel went so far as to suggest that perhaps proof was impossible, and he remarked that our certainty concerning this postulate rests not on the possibility of rigorous deduction, nor in the definition of straight lines, but on empirical grounds and the evidence of sensory perceptions.

The work was read by the Swiss mathematician Johann Lambert (1728-1777), who wrote a monograph on the "Theory of Parallels" in 1766. Lambert adopted an approach somewhat similar to that of Saccheri, whose work he no doubt had seen in Klügel's treatise.

Lambert considered a trirectangular quadrilateral ABCD with right angles at A, B, and C (and no assumption concerning the lengths of the sides). For the fourth angle, ∡D, Lambert framed three hypotheses, analogous to those of Saccheri.

The hypothesis of the right angle led to Euclidean geometry. The hypothesis of the obtuse angle quickly led to a contradiction. Assuming the hypothesis of the acute angle, Lambert made some astonishing discoveries of its implications, in particular, analogies between the resulting geometry and that of a sphere.

He discovered that in such a geometry the angle sum of a triangle would be less than π and that the area of a triangle would be proportional to its defect. (This should come as no surprise to us, inasmuch as both area and defect are additive—see Exercise 2 of Section 2.) He showed also that if a geometry based on the obtuse angle hypothesis were possible, then a triangle's angle sum would exceed π and the *excess* would be proportional to the area. The latter is exactly what happens in spherical geometry, where the area of a triangle $\triangle ABC$ is given by

$$\text{Area} = r^2 (\angle A + \angle B + \angle C - \pi)$$

where r is the radius of the sphere and the angles are measured in radians. Of course, spherical geometry is inconsistent with the postulates of absolute geometry.

Since the triangle area formula in the acute angle geometry could be obtained from the spherical formula (above) by formally replacing r by $(-1)^{\frac{1}{2}} r$, Lambert was prompted to remark that apparently the third hypothesis would occur on a sphere with "imaginary radius" (whatever that might mean).*

Because the sides of an equilateral triangle are determined by its defect and so by its angles, Lambert found that lengths, as well as angles, would have absolute, rather than relative meaning in the geometry of the acute angle hypothesis (cf. Section 3).

Although the existence of absolute units of length seemed outlandish to him, nevertheless Lambert was astute enough not to think that from this he had proved the necessity of Euclid's postulate. He pressed on with his investigations, but like Saccheri before him, was unable to reach his goal. Lambert did not see fit to publish his work himself. His *Theory of Parallels* was published posthumously.

---

*How prophetic this remark was will be seen in Chapter VI.

Exercises IV-5

1. Show that if the angle sum of every triangle is less than π, then the fourth angle in every Lambert quadrilateral is acute.

2. Let ABCD be a Lambert quadrilateral with acute angle at A and vertex C opposite A. In Lobachevskian geometry, which is longer, $\overline{AB}$ or $\overline{CD}$? Prove your answer.

3. Let ABCD and A'B'C'D' be Lambert quadrilaterals with acute angles at A and A'. Prove that if $\angle A \cong \angle A'$ and AB = A'B', then the quadrilaterals are congruent (corresponding parts are congruent.)

4. Prove that the line joining the midpoints of the two equal sides of a Saccheri quadrilateral is perpendicular to the line joining the midpoints of the base and the summit, and that it bisects the diagonals. (In the figure below, show $\overleftrightarrow{EF} \perp \overleftrightarrow{MN}$, AH = CH, and BG = DG. You should not assume that $\overleftrightarrow{EF}$, $\overleftrightarrow{MN}$, and the two diagonals are all concurrent. In fact, that cannot happen in Lobachevskian geometry, for if H = P, then PFMB would be a Lambert quadrilateral, and by Exercise 2, FH > MG = $\frac{1}{2}$ AB, contradicting Exercise 5 of Section 4. You can show, however, that $\overleftrightarrow{MN}$ and the diagonals are concurrent.)

Definition. Two polygons p and q are called *equivalent* if they can be partitioned into the same finite number of triangles, $p = p_1 \cup p_2 \cup \ldots \cup p_n$, $q = q_1 \cup q_2 \cup \ldots \cup q_n$ (with disjoint interiors) such that $p_i \cong q_i$ for all i. In the following exercises, assume equivalence in this sense is an equivalence relation (see Wolfe [53, pp. 122ff]).

Definition. The *defect* of a polygon of n sides is the difference between $(n-2)\pi$ and the sum of the polygon's interior angles. Assume that when a polygon is partitioned into a finite number of subpolygons, the defect of the whole is the sum of the defects of the parts.

5. Prove that a given triangle is equivalent to a Saccheri quadri-

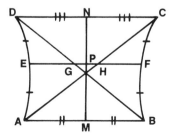

lateral whose summit is one of the triangle's sides and whose defect equals that of the triangle.

6. Show that two triangles which have the same defect and a side of one congruent to a side of the other are equivalent in Lobachevskian geometry. (Hint: cf. Exercise 10, Section IV-4.)

7. Let Saccheri quadrilateral ABGF be constructed from △ABC as in Figure IV-14. Suppose (see the figure below) that E″ is any point on line $\overleftrightarrow{DE}$. Extend $\overline{BE''}$ through E″ to C″ so that BE″ = E″C″. Show that triangles △ABC and △ABC″ have the same defect and are equivalent.

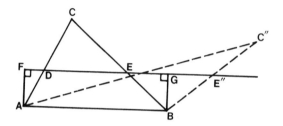

(Note that this generalizes the result that the triangles derived from △ABC share its defect; cf. the proof of Theorem IV-2.)

8. Show that two triangles with the same defect are equivalent. (Hint: if a side of one triangle is congruent to a side of the other, we are done, by Exercise 6. Therefore assume, in triangles △ABC and △A′B′C′ that B′C′ > BC. Construct a Saccheri quadrilateral associated with △ABC as in Figure IV-14. In Exercise 7, choose E″ so that BE″ = $\frac{1}{2}$B′C′—explain why this is possible. Show that △ABC″ and △A′B′C′ fit the hypotheses of Exercise 6.)

Note on areas. Because squares do not exist in Lobachevskian geometry, area cannot be defined in terms of "square units." However, since two triangles are equivalent if and only if they have the same defect, it seems reasonable to define the *area* of a triangle, or more generally, the *area* of a polygon to be a constant times its defect. The value of this constant can be fixed by choosing some convenient polygon and declaring it to have unit area. Alternatively, for given choices of the unit of distance and the unit of angle measure, the constant is uniquely determined if we require that the area formulas of Lobachevskian geometry approach their Euclidean counterparts as the lengths of the sides of figures approach zero (this will be made clearer in Chapter VI). In the following problems, assume angles are measured in radians, and denote the constant described in the previous sentence by k^2. Then, for any polygon,

Area = k^2 × Defect

A similar situation pertains on the sphere. If angles are measured in radians, the area of a spherical polygon is given by Area = R^2 × Excess, where R is the sphere's radius. The numerical value of R depends of course on the choice of unit of distance. In Lobachevskian and spherical geometry then, the unit of area depends on both the unit of distance and the unit of angle measure. In Euclidean geometry, it depends only on the former.

9. (a) What is the area of a Saccheri quadrilateral with summit angle α (radians)? (b) What is the area of a Lambert quadrilateral with acute angle α?

10. Prove that two Saccheri quadrilaterals with the same area and congruent bases are congruent.

11. (a) Find the area of an equilateral triangle each of whose angles is α. (b) What is the least upper bound of the areas of equilateral triangles in Lobachevskian geometry?

12. In Lobachevskian geometry, how large can the interior angles of a regular n-gon (a polygon with n congruent sides and n congruent interior angles) be? How small?

6. THE FRENCH GEOMETERS

Toward the end of the 18th century, many of the great French mathematicians wrote on the subject of parallels and attempted to derive the Euclidean postulates.

D'Alembert (1717-1783), known chiefly for his work in the science of mechanics, believed that the difficulties could be surmounted by means of the right definition of a straight line. He suggested defining a parallel as a coplanar line connecting two points which are on the same side of and equidistant from a given line. The problem—equivalent to proving P5—was then to show that parallel lines are everywhere equidistant, not just at the two points.

Lagrange (1736-1813) thought he had succeeded, but in the midst of presenting a paper on parallels before the French Academy, interrupted his reading with the exclamation "I must think on this some more."

Laplace (1749-1827), famous for his work in the fields of celestial mechanics and probability theory, linked the parallel question to Newton's Law of Gravitation. Like many others before him, he approached the problem through the existence of similar figures (see Theorem IV-8), his argument being that if all lengths, distances, and velocities in the universe were shrunk in the same proportion, then all celestial bodies would follow paths precisely similar to those they follow now. In other words, there are no absolute dimensions to the universe, only ratios of dimensions. "This proportionality," he declared, "appears to me a more natural postulate than that of Euclid, and it is worthy of note that it is discovered afresh in the results of the theory of universal gravitation."

Adrien Marie Legendre (1752-1833) published several attempts to prove the parallel postulate in successive editions of his *Elements of Geometry*, between 1794 and 1823. Although his results were not new, his elegant and lucid style gained him a large following and no doubt helped pave the way for the discovery of non-Euclidean geometry. Although proved by Saccheri, Theorems IV-2 and IV-5 are often called Legendre's first and second theorems, and the proofs we have given are his.

In each of his several attempts to prove the parallel postulate, Legendre made use of some assumption, tacitly or otherwise, that is equivalent to the postulate. For example, in one "proof," he resorts to the concept of similarity in the assumption that lengths are relative. In another, he assumes that through three non-collinear points passes a circle. This also turns out to be equivalent to P5 (see Section 7).

In one attempt, Legendre assumes the existence of a triangle, $\triangle ABC$, with defect $\delta > 0$, and proceeds toward a "contradiction" as follows. Reflect $\triangle ABC$ in side $\overline{BC}$, i.e., adjoin to $\triangle ABC$ a congruent triangle, $\triangle BCD$, in which $BD = AB$ and $CD = AC$ (Fig. IV-15). Through point D draw any line which meets the extensions of sides $\overline{AB}$ and $\overline{AC}$. Call the intersections E and F, respectively. The resulting triangle, $\triangle AEF$, is composed of four triangles, two of which are congruent to $\triangle ABC$. Then, because defects add, $\triangle AEF$ has defect $> 2\delta$.

If we repeat the construction (reflecting $\triangle AEF$ in side $\overline{EF}$, etc.), we will obtain a triangle with defect $> 4\delta$. After n steps, we obtain a

THE FRENCH GEOMETERS

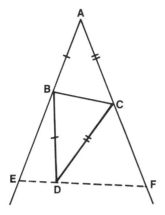

Figure IV - 15.

triangle with defect $> 2^n \delta$, which, for sufficiently large n, exceeds π. This is impossible since, in any triangle $D = \pi - \mathscr{S} < \pi$. Therefore, there can be no triangle with a positive defect, and so Euclid's postulate must hold.

In the above attempt to prove the parallel postulate, Legendre is unwittingly assuming something equivalent to it. Can you guess what it is? (Legendre's unwarranted assumption will be revealed in Chapter V.)

Exercises IV-6

1. In the figure, $\overline{AB}$ is a diameter of the circle. Prove $\angle C = \frac{1}{2} \mathscr{S}$ (ABC). (Hint: draw radius $\overline{OC}$.)

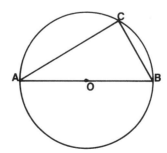

2. Prove that in Lobachevskian geometry the angles inscribed in a semicircle are not all congruent. (Hint: for each of the inscribed triangles in the figure, we may construct Saccheri quadrilaterals as in

Fig. IV-14. Show that ∡C = ∡D would imply these quadrilaterals are congruent. This is clearly impossible for all choices of C and D.)

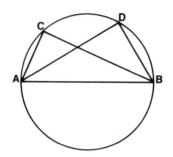

7. WOLFGANG BOLYAI

The Hungarian mathematician *Wolfgang Bolyai* (1775-1856) became interested in the theory of parallels during his student days with Gauss at Göttingen, probably under the encouragement of Kästner, the Professor of Mathematics and the mentor of G. S. Klügel (Section 5).

In 1804, Bolyai sent to Gauss a "Theory of Parallels" in which he claimed to have proved the existence of equidistant straight lines. Gauss pointed out the flaw in Bolyai's argument. Four years later, Bolyai sent another note to Gauss, but the latter failed to reply.

Some years later, Wolfgang collected his thoughts on the subject in his publication *Tentamen juventutem studiosam in elementa matheseos* ("Essay for studious youths on the elements of mathematics," 1832-33). This work's chief claim to fame is that it contained a now famous *Appendix*, written by Wolfgang's son, Johann, one of the co-discoverers of non-Euclidean geometry (Section 9). See [5].

Wolfgang discovered that a statement surprisingly equivalent to the parallel postulate is "a circle can always be drawn though three points not on a straight line." (We have already noted that Legendre assumed this proposition in one of his "proofs" of the parallel postulate.) Any student of elementary geometry is aware that this proposition holds in Euclidean geometry. To prove, conversely, that it implies P5 (or equivalently, Playfair's Axiom), let P be a point not on a given line ℓ, and let Q be the foot of the perpendicular from P to ℓ. Suppose $\overleftrightarrow{PA}$ is any line through P for which ∡APQ is acute (Fig. IV-16). We shall show that $\overleftrightarrow{PA}$ meets ℓ.

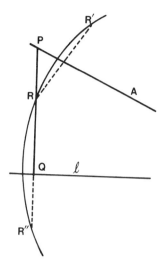

Figure IV - 16.

Let R be a point on segment $\overline{PQ}$ and let R' and R" be its reflections in $\overleftrightarrow{PA}$ and ℓ, respectively. Specifically, $\overleftrightarrow{PA}$ is the perpendicular bisector of $\overline{RR'}$ and ℓ is the perpendicular bisector of $\overline{RR''}$. Because ⊀APQ is acute, R, R', and R" are not collinear. Therefore, by our hypothesis, there is a circle passing through these three points. But $\overleftrightarrow{PA}$ and ℓ, being perpendicular bisectors of chords, are diameters of this circle. They must therefore intersect at the circle's center. Hence, of all the lines through P, only the perpendicular to PQ fails to meet ℓ. We have verified Playfair's Axiom.

8. GAUSS

Perhaps the first person to fully comprehend the geometry independent of Euclid's fifth postulate was *Carl Friedrich Gauss* (1777-1855) the dominant mathematical figure of his time, and undoubtedly one of the greatest mathematicians of all time. His meditations on the subject can be traced through letters written to colleagues over a period of three decades.

Initially, like Saccheri and Lambert, Gauss tried to prove the parallel postulate by deducing a contradiction from its negation. He believed Euclidean geometry was the geometry of physical space, although he most likely would have admitted the possibility of other logically consistent geometries. In 1799, he wrote to Wolfgang Bolyai, who at that time had thought he had solved the problem [5].

> As for me, I have already made some progress in my work. However the path I have chosen does not lead at all to the goal which we seek, and which you assure me you have reached. It seems rather to compel me to doubt the truth of geometry itself.
> It is true that I have come upon much which by most people would be held to constitute a proof: but in my eyes it proves as good as nothing. For example, if one could show that a rectilinear triangle is possible, whose area would be greater than any given area, then I would be ready to prove the whole of geometry absolutely rigorously.
> Most people would certainly let this stand as an Axiom; but I, no! It would, indeed, be possible that the area might always remain below a certain limit, however far apart the three angular points of the triangle were taken.

As Gauss suspected, there is indeed an upper bound to the areas of triangles in Lobachevskian geometry. This is imlicit in Lambert's discovery that the area of a triangle is proportional to its defect (π is an upper bound for defects).

Before long, Gauss became convinced that Euclid's parallel postulate was independent of the other assumptions of geometry. After about 1813, he had come to accept the existence of a new geometry in which more than one parallel to a given line can be drawn through a given external point. He calls this geometry first *Anti-Euclidean,* then *Astral Geometry,* and finally *non-Euclidean Geometry.* He considered whether this new geometry might be suitable for the description of physical space. In 1817, writing to his close friend, the noted amateur astronomer Wilhelm Olbers (a practicing physician by day), he laments

> I keep coming closer to the conviction that the necessary truth of our geometry cannot be proved, at least by the human intellect for the human intellect. Perhaps in another life we shall arrive at other insights into the nature of space which at present we cannot reach. Until then we must place geometry on an equal basis, not with arithmetic, which has a purely *a priori* foundation, but with mechanics.

By "the necessary truth of our geometry," Gauss means "the physical necessity of Euclidean geometry." Gauss is labeling geometry as an empirical science.

Around this time, Gauss was employed by the Hanoverian government to participate in a geodetic survey. His experience in surveying and map making naturally stimulated his interest in applied geometry, in particular the study of curved surfaces, to which he contributed a landmark paper, *General Investigations of Curved Surfaces,* in 1827 [14].

Using light signals, Gauss actually measured the sum of the angles in a triangle formed by three mountain peaks. The sides of the triangle were on the order of fifty miles in length. He found the sum to be 180°, as far as the accuracy of his instruments could tell. Calculating the angles of the corresponding geodesic (great circle) triangle on the earth's surface, he arrived at a surprisingly accurate estimate of the excess of that triangle's angle sum over 180° (see Exercise 1). He remarked that the flattening of the earth at the poles could be neglected in the measurement of all such triangles on the earth's surface. These observations may have had as a secondary purpose the testing of the Euclidean hypothesis in three-dimensional physical space, although Gauss did not explicitly state this (see Hall, [17] p. 123ff.). Because of inevitable observational errors, Gauss's triangle measurements shed no light on this latter question.

Since the defect of a triangle is proportional to its area, we would expect that the measurement of much larger triangles, perhaps triangles of astronomical dimensions, would be more likely to shed light on the question of which geometry describes physical space best. Such measurements were made by Lobachevsky (cf. Chapter VIII).

The suggestion that physical space might be something other than Euclidean ran counter to the philosophy prevalent in the early 19th century. Immanuel Kant (1724-1804), in his *Critique of Pure Reason* (1781) held that knowledge of space was not empirical but intuitive, stemming from concepts existing *a priori* in the human mind. The influx of sensory perceptions impinging on our consciousness is given form and order by our minds, just as a fluid is given shape by the vessel into which it is poured. In all of us, these perceptions are always organized or formed according to a Euclidean framework, i.e., within a Euclidean context. Since we are all born with a Euclidean intuition of space, there can be only one kind of geometry. Scientific and mathematical discoveries since the time of Kant have largely invalidated this philosophy. Where Kant held that the idea of space is something forced upon us by our consciousness, and therefore independent of the presence or absence of material objects, Einstein's general theory of relativity teaches us that the geometric properties of space are determined by the distribution of matter in space. Nevertheless, Kant's philosophy had a powerful influence on the intellectual community, and indeed was at least partly responsible for the long delay in the acceptance of non-Euclidean geometry.

By 1824, Gauss had worked out the details of his non-Euclidean geometry, as we gather from this letter to Taurinus [53, pp. 46-47].

> In regard to your attempt, I have nothing (or not much) to say except that it is incomplete. It is true that your demonstration of the proof that the sum of the three angles of a plane triangle cannot be greater than 180° is somewhat lacking in geometrical rigor. But this in itself can easily be remedied, and there is no doubt that the impossibility can be proved most rigorously. But the situation is quite different in the second part, that the sum of the angles cannot be less than 180°; this is the critical point, the reef on which all the wrecks occur. I imagine that this problem has not engaged you very long. I have pondered it for over thirty years, and I do not believe that anyone can have given more thought to this second part than I, though I have never published anything on it. The assumption that the sum of the three angles is less than 180° leads to a curious geometry, quite different from ours (the Euclidean), but thoroughly consistent, which I have developed to my entire satisfaction, so that I can solve every problem in it with the exception of the determination of a constant, which cannot be designated a priori. The greater one takes this constant, the nearer one comes to Euclidean Geometry, and when it is chosen infinitely large the two coincide. The theorems of this geometry appear to be paradoxical and, to the uninitiated, absurd; but calm, steady reflection reveals that they contain nothing at all impossible. For example, the three angles of a triangle become as small as one wishes, if only the sides are taken large enough; yet the area of the triangle can never exceed a definite limit, regardless of how great the sides are taken, nor indeed can it ever reach it. All my efforts to discover a contradiction, an inconsistency, in this Non-Euclidean Geometry have been without success, and the one thing in it which is opposed to our conceptions is that, if it were true, there must exist in space a linear magnitude, determined for itself (but unknown to us). But it seems to me that we know, despite the say-nothing word-wisdom of the metaphysicians, too little, or too nearly nothing at all, about the true nature of space, to consider as absolutely impossible that which appears to us unnatural. If this Non-Euclidean Geometry were true, and it were possible to compare that constant with such magnitudes as we encounter in our measurements on the earth and in the heavens, it could then be determined a posteriori. Consequently in jest I have sometimes expressed the wish that the Euclidean Geometry were not true, since then we would have a priori an absolute standard of measure.
>
> I do not fear that any man who has shown that he possesses a thoughtful mathematical mind will misunderstand what has been said above, but in any case consider it a private communication of which no public use or use leading in any way to publicity is to be made. Perhaps I shall myself, if I have at some future time more leisure than in my present circumstances, make public my investigations.

As we shall see in Chapter V, the "constant" to which Gauss refers is the *space constant* or *radius of curvature* of the Lobachevskian plane and is the analogue of the radius of the sphere in spherical geometry. The "linear magnitude, determined for itself" alludes to the fact that lengths are absolute in Lobachevskian geometry. The "say-nothing word-wisdom of the metaphysicians" is likely meant as a slap at Kant and his followers.

It is significant that Gauss, whose mathematical reputation was unassailable, considered the new geometry so revolutionary that he withheld publication. In 1829 he wrote, in a private communication to Bessel,

> It may take very long before I make public my investigations on this issue; in fact, this may not happen in my lifetime for I fear the "clamor of the Boeotians."

Boeotia was a province of ancient Greece whose inhabitants were known for their dullness and ignorance.

It is well-known that Gauss refrained from publishing many of his greatest discoveries. He had a strong aversion to controversy and public debate, which surely would have been precipitated by early publication of his work on imaginary numbers and the new geometry. Moreover, he preferred to refine his proofs and polish his expositions over and over until a complete and flawless masterpiece emerged, with no trace of earlier false starts or less perfect alternative paths. His seal showed a tree with but seven fruits and the motto "few, but ripe." Had Gauss published his work on non-Euclidean geometry in 1824, history would have credited him with the discovery of the new geometry. Instead, the credit went to those who published first. Eventually, Gauss did make the decision to publish. In 1831, he wrote to Schumacher [53, p. 48],

> I have begun to write down during the last few weeks some of my own meditations, a part of which I have never previously put in writing, so that already I have had to think it all through anew three or four times. But I wished this not to perish with me.

However, before completing this task, Gauss received a copy of J. Bolyai's Appendix.

Exercises IV-8

1. The sides of the mountain peak triangle which Gauss measured are approximately 69 km, 85 km, and 107 km. As you can verify, this is almost a right triangle. The radius R of the earth is approximately 6371 km. Using the spherical geometry area formula

$$\text{Area} = R^2 \times \text{Excess}$$

(where the excess is in radians), estimate the excess of that spherical triangle. Why is it permissible to estimate the triangle's area by means of Euclidean geometry, even though its sides are great circle arcs?

9. J. BOLYAI

Johann (or *Janos*) *Bolyai* (1802-1860), the son of Wolfgang, displayed great talent for mathematics as a youth. When only 13, he would occasionally lecture in his father's stead. The students said they prefered being taught by the son. He was already an accomplished violinist. At fifteen, he entered the Engineering Academy in Vienna, where he devoted much of his time in trying to find a proof of Euclid's postulate. In 1823, he was appointed to the military.

Endowed with a haughty temper, he once accepted challenges from 13 cavalry officers, on the condition that he be allowed to play a short piece on the violin after each duel. He overcame all 13 adversaries.

Wolfgang had actually admonished his son to leave the matter of parallels aside [31, p. 31]:

> You must not attempt this approach to parallels. I know this way to its very end. I have traversed this bottomless night, which extinguished all light and joy of my life. I entreat you, leave this science of parallels alone . . .

Nevertheless, Johann continued his investigations and by 1823 had worked out an "absolute theory" of geometry in which he presupposed neither Euclid's postulate nor its negation. For example, he proves that

> In any rectilinear triangle, the [circumferences of the] circles with radii equal to its sides are [in proportion] as the sines of the opposite sides.

If we denote the circumference of the circle with side x by Circ(x), then the above result says that if, in a triangle, a, b, and c are the sides opposite angles A, B, and C, respectively, then

$$\text{Circ}(a)/\sin A = \text{Circ}(b)/\sin B = \text{Circ}(c)/\sin C$$

As stated, this theorem is *absolute* and holds no matter which parallel postulate is adopted. In Euclidean geometry, Circ(x) = $2\pi x$ and the formula becomes the familiar Law of Sines,

$$a/\sin A = b/\sin B = c/\sin C$$

Bolyai showed that under the negation of P5, Circ (x) = $2\pi k \sinh(x/k)$, where k is an undetermined constant. (The function sinh is defined in Appendix C.) This gives

$$\sinh(a/k)/\sin A = \sinh(b/k)/\sin B = \sinh(c/k)/\sin C$$

These relations will be derived in Chapter VI.
In 1823, Johann wrote his father [5, p. 98],

> I have now resolved to publish a work on the theory of parallels, as soon as I shall have put the material in order, and my circumstances allow it. I have not yet completed this work, but the road which I have followed has made it almost certain that the goal will be attained, if that is at all possible: the goal is not yet reached, but I have made such wonderful discoveries that I have been almost overwhelmed by them, and it would be the cause of continual regret if they were lost. When you will see them, you too will recognize it. In the meantime I can say only this: I have created a new universe from nothing. All that I have sent you till now is but a house of cards compared to the tower. I am as fully persuaded that it will bring me honour, as if I had already completed the discovery.

Wolfgang urged his son to make known his discoveries as soon as possible [5, p. 99]:

> If you have really succeeded in the question, it is right that no time be lost in making it public, for two reasons: first, because ideas pass easily from one to another, who can anticipate its publication; and secondly, there is some truth in this, that many things have an epoch, in which they are found at the same time in several places, just as the violets appear on every side in spring. Also every scientific struggle is just a serious war, in which I cannot say when peace will arrive. Thus we ought to conquer when we are able, since the advantage is always to the first comer.

In 1829, Johann sent his manuscript to his father, who agreed that it should appear as an appendix to his *Tentamen,* which was published in 1832. Johann's treatise bore the Latin title *Appendix Scientiam Spatii absolute veram exhibens: a veritate aut falsitate Axiomatis XI Euclidei* (*a priori haud unquam decidenda*) *independentem* ("Appendix exhibiting the absolute science of space: independent of the truth or falsity of Euclid's Axiom XI (by no means previously decided)"). Euclid's parallel postulate was listed as Axiom XI in several 16th and 17th century editions of the *Elements* (see Bonola [5, pp. 17-21] for a discussion of the original distinction between the words "axiom" and "postulate," now used interchangeably).

A copy of the Appendix was sent to Gauss, who replied in a manner that was somewhat disturbing to the young Johann [5, p. 100].

> If I commenced by saying that I am unable to praise this work, you would certainly be surprised for a moment. But I cannot say otherwise. To praise it, would be to praise myself. Indeed the whole contents of the work, the path taken by your son, the results to which he is led, coincide almost entirely with my meditations, which

> have occupied my mind partly for the last thirty or thirty-five years. So I remained quite stupefied. So far as my own work is concerned, of which up till now I have put little on paper, my intention was not to let it be published during my lifetime. Indeed the majority of people have not clear ideas upon the questions of which we are speaking, and I have found very few people who could regard with any special interest what I communicated to them on this subject. To be able to take such an interest it is first of all necessary to have devoted careful thought to the real nature of what is wanted and upon this matter almost all are most uncertain. On the other hand it was my idea to write down all this later so that at least it should not perish with me. It is therefore a pleasant surprise for me that I am spared this trouble, and I am very glad that it is just the son of my old friend, who takes the precedence of me in such a remarkable manner.

Johann could not escape the feeling that Gauss had learned of his discoveries from the elder Bolyai prior to the publication of the Appendix and that Gauss was trying to claim priority unjustly. He later came to realize that his suspicions had been unfounded. It was many years before the work of Johann Bolyai gained the recognition it deserved. Even Gauss, who heartily approved of this work, hesitated to endorse it in print. Johann published nothing further.

10. LOBACHEVSKY

Credit for the first published development of non-Euclidean geometry as a logically consistent system goes to the Russian mathematician *Nicolai Ivanovitch Lobachevsky* (1793-1856). Lobachevsky studied at the University of Kasan, where he became Professor and then Rector in 1827. About 1815, he began work at the theory of parallels, attempting to prove Euclid's fifth postulate by an approach similar to that of Legendre. By about 1823, he had given up these attempts and turned to the development of an alternative geometry where two intersecting lines can be parallel to a third line and where the sum of the angles in a triangle is less than two right angles. Lobachevsky called this geometry *Imaginary Geometry*, and later, *Pangeometry*.

Like Gauss, Lobachevsky believed that geometry was an empirical science [5, p. 92] :

> The fruitlessness of the attempts made, since Euclid's time, for the space of 2000 years, aroused in me the suspicion that the truth, which it was desired to prove, was not contained in the data themselves; that to establish it the aid of experiment would be needed, for example, of astronomical observations, as in the case of other laws of nature. When I had finally convinced myself of the justice of my conjecture and believed that I had completely solved this difficult question, I wrote, in 1826, a memoir on this subject.

This memoir was presented in 1826 to the mathematics and physics department of the University of Kasan in the form of a lecture (in French) entitled *A succinct exposition of the principles of geometry with a rigorous demonstration of the theorem of parallels.* The manuscript for this lecture was lost.

However, in 1829, he published a second memoir in the Kasan Bulletin, *On the Principles of Geometry.* This was the earliest published account of non-Euclidean geometry, which is now known as Lobachevskian or hyperbolic geometry (for reasons which will become clearer later on).

In latter years, Lobachevsky published a number of expositions, the most important of which was *Geometric Researches on the Theory of Parallels* (1840), written in German, and appearing in English translation in Bonola [5]. Lobachevsky's work is strikingly similar to that of J. Bolyai—and indeed to that of Gauss, although the evidence indicates that the three worked independently—but Lobachevsky gave special emphasis to the non-Euclidean trigonometric formulas and the analytical form of his geometry. Referring to a later exposition published in 1855, he wrote [5, pp. 93-94]:

> Now that we have shown, in what precedes, the way in which the lengths of curves, and the surfaces and volumes of solids can be calculated, we are able to assert that the Pangeometry is a complete system of geometry. A single glance at the equations which express the relations existing between the sides and angles of plane triangles, is sufficient to show that, setting out from them, Pangeometry become a branch of analysis, including and extending the analytical methods of ordinary geometry.

What he is saying here is that the success of the analytic formulation of his "pangeometry" indicates (but does not conclusively prove of course) that this geometry is consistent.

Lobachevsky not only saw the connection between his geometry and spherical, but also discovered that for small triangles the non-Euclidean trigonometric formulas agree, to a first approximation, with the formulas of ordinary (Euclidean) trigonometry. This is very significant for the application of geometry to physical space. As we shall see in Chapter VIII, Lobachevsky made measurements in an effort to discover whether physical space is non-Euclidean.

By introducing coordinates, Lobachevsky was able to calculate the lengths of curves, areas of surfaces, and volumes of certain solids in non-Euclidean geometry. In calculating the volume of a solid in different ways, Lobachevsky obtained relations between different types of definite integrals—an application of "imaginary" geometry to real analysis.

Gauss learned of Lobachevsky's work from the 1840 *Geometric Researches*. In 1846, he wrote [53, p. 55]

> I have recently had occasion to look through again that little volume by Lobatschefski* (Geometric Researches on the Theory of Parallels). It contains the elements of that geometry which must hold, and can with strict consistency hold, if the Euclidean is not true. A certain Schweikardt calls such geometry Astral Geometry, Lobatschefski calls it Imaginary Geometry. You know that for fifty-four years now (since 1792) I have held the same conviction (with a certain later extension, which I will not mention here). I have found in Lobatschefski's work nothing that is new to me, but the development is made in a way different from that which I have followed, and certainly by Lobatschefski in a skillful way and in truly geometrical spirit. I feel that I must call your attention to the book, which will quite certainly afford you the keenest pleasure.

Gauss's claim that he had believed in a consistent non-Euclidean geometry as early as 1792 is perhaps an exaggeration.

From Gauss, Wolfgang and then Johann became acquainted with the *Geometric Researches* of Lobachevsky in 1848. Johann had this to say [53, p. 56]:

> Even if in this remarkable work different methods are followed at times, nevertheless, the spirit and result are so much like those of the *Appendix* to the *Tentamen matheseos* which appeared in the year 1832 in Maros-Vásárhely, that one cannot recognize it without wonder. If Gauss was, as he says, surprised to the extreme, first by the *Appendix* and later by the striking agreement of the Hungarian and Russian mathematicians: truly, none the less so am I.
>
>   The nature of real truth of course cannot but be one and the same in Maros-Vásárhely as in Kamschatka and on the Moon, or, to be brief, anywhere in the world; and what one finite, sensible being discovers, can also not impossibly be discovered by another.

Although the honor goes to Lobachevsky and to a lesser extent Bolyai (and Gauss), the discovery of non-Euclidean geometry, far from being a spontaneous flash of enlightenment, was brought about gradually by a number of people over a long period of time. The achievements of Lobachevsky and Bolyai were the culmination of over two millenia of striving toward an understanding of the nature of geometric truth. Men like Lambert, Legendre, Schweikardt,

---

*Being a transliteration from the Russian, Lobachevsky's name is spelt in several ways.

Taurinus, as well as many others whom we have not mentioned for lack of space, also played an important role. These men contributed to a revolution in geometry that has been compared to the 16th century Copernican revolution in astronomy. In the remainder of this book, we shall examine the content of non-Euclidean geometry, its subsequent development, and its implications for the study of the physical universe.

V

FUNDAMENTALS OF LOBACHEVSKIAN GEOMETRY

> *Mathematicians are like lovers... Grant a mathematician the least principle and he will draw from it a consequence which you must also grant him, and from this consequence another.*
>
> —Fontenelle

1. PARALLELISM OF RAYS

In Lobachevskian geometry, we retain all of Euclid's postulates, or rather, all of Hilbert's postulates, except the parallel postulate, which Lobachevsky replaces with the following.

Lobachevsky's Parallel Postulate

Through a point not on a given line there exists (in the plane determined by this point and line) at least two lines which do not meet the given line (Fig. V-1).

Note that we use the phrase "do not meet" rather than "are parallel to." The word "parallel," as we shall see below, is not given the same definition in Lobachevsky's geometry as in Euclid's.

Definition V-1

We shall call ray $\overrightarrow{PR}$ *parallel* to ray $\overrightarrow{QS}$ if they do not meet, and if every ray interior to ∡RPQ meets $\overrightarrow{QS}$. We write $\overrightarrow{PR} \parallel \overrightarrow{QS}$.

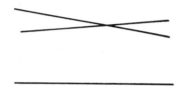

Figure V - 1.

It is easy to see (Fig. V-2) that there is at most one ray through P which is parallel to $\overrightarrow{QS}$ (if you draw a sketch with two and reread the definition, you will see why). Moreover, the sum of the interior angles, ∡QPR and ∡PQS, cannot exceed two right angles, by Euclid I,28, which is a theorem of absolute geometry. (The sum equals π in Euclidean geometry, but, as we shall see later from Theorem V-16, is less than π in Lobachevskian geometry.)

Note: In Lobachevskian geometry, we may freely apply any theorems of Euclid which are absolute, as long as, whenever Euclid uses the word *parallel* we interpret him to mean "not meeting." It is important to remember this distinction.

To verify the existence of a ray through P which is parallel to $\overrightarrow{QS}$, consider the angle θ which a ray $\overrightarrow{PR}$ makes with $\overline{PQ}$ (Fig. V-3). We may think of $\overrightarrow{PR}$ as free to pivot about P, so that θ is between zero and let's say, a straight angle. Let

$\mathscr{X} = \{ \theta$ such that $\overrightarrow{PR}$ meets $\overrightarrow{QS} \}$
$\mathscr{Y} = \{ \theta$ such that $\overrightarrow{PR}$ does not meet $\overrightarrow{QS} \}$.

By joining P to any point on $\overrightarrow{QS}$, we see that $\mathscr{X}$ is not empty. $\mathscr{Y}$ is not empty because it contains π - ∡PQS, by Euclid I,28. Furthermore, each angle of $\mathscr{X}$ is smaller than every angle of $\mathscr{Y}$, as the opposite would contradict Theorem III-17 (check this).

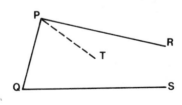

Figure V - 2.

PARALLELISM OF RAYS

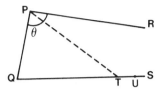

Figure V - 3.

It then follows from Dedekind's postulate that there is a unique angle which forms the "border" between $\mathscr{X}$ and $\mathscr{Y}$. The corresponding ray $\overrightarrow{PR_o}$ does not meet $\overrightarrow{QS}$, for otherwise, calling the intersection T and joining P to a point U such that T is between Q and U, we would obtain an angle, ∡UPQ, in $\mathscr{X}$, larger than ∡TPQ (=∡R_oPQ), a contradiction. Since rays interior to ∡R_oPQ meet QS, $\overrightarrow{PR_o}$ ∥ $\overrightarrow{QS}$.

2. ANGLE OF PARALLELISM

An important case arises when the "transversal" $\overleftrightarrow{PQ}$ is perpendicular to $\overleftrightarrow{QS}$. If P is a point not on a given line $\overleftrightarrow{AA'}$, let Q be the foot of the perpendicular from P to $\overleftrightarrow{AA'}$. Through P there exist two rays $\overrightarrow{PR}$ and $\overrightarrow{PS}$ parallel, respectively, to rays $\overrightarrow{QA'}$ and $\overrightarrow{QA}$ (Fig. V-4). Then we claim ∡SPQ ≅ ∡RPQ. For, if one of them, say ∡SPQ, were smaller, we could draw, within the larger ∡RPQ, ray $\overrightarrow{PT}$ such that ∡TPQ ≅ ∡SPQ. By parallelism, $\overrightarrow{PT}$ would meet $\overrightarrow{QA'}$ in a point, which we may as well label T to conserve on symbolism (a ploy used throughout this book). On $\overrightarrow{QA}$ mark U so that QU = QT. The congruence of △PQT and △PQU yields ∡TPQ ≅ ∡UPQ < ∡SPQ ≅ ∡TPQ, a contradiction.

Definition V-2

The angle ∡SPQ (≅ ∡RPQ) is called the *angle of parallelism* and is denoted Π(PQ).

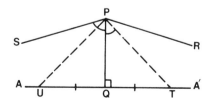

Figure V - 4.

We shall see shortly that Π(PQ) depends only on the distance from P to $\overleftrightarrow{AA'}$.

In Euclidean geometry (by Playfair's postulate), the lines determined by the rays $\overrightarrow{PR}$ and $\overrightarrow{PS}$ must coincide, and so Π(PQ) is always a right angle. However, under Lobachevsky's postulate (which we henceforth assume, even though many of the results to follow are valid in Euclidean geometry as well), these lines are necessarily distinct and Π(PQ) is acute.

The following results tell us that parallelism of rays depends only on the lines determined by the rays and not on their initial points.

Lemma V-3

If rays $\overrightarrow{PR} \parallel \overrightarrow{QS}$, then rays $\overrightarrow{AR} \parallel \overrightarrow{QS}$ for any A on line $\overleftrightarrow{PR}$. [It is assumed here that R has been chosen far enough away from P so that R is not between P and A (Fig. V-5a). There is no loss of generality, since the definition of ray $\overrightarrow{PR}$ does not depend on the choice of point R (see Exercise 9 of Section III-3).]

Proof: If A lies on ray $\overrightarrow{PR}$, and if B is a point in the interior of figure RAQS, then ray $\overrightarrow{PB}$, by the parallelism of $\overrightarrow{PR}$ with $\overrightarrow{QS}$, must meet $\overrightarrow{QS}$ in a point C. Then Pasch's axiom, applied to △PQC, implies that ray $\overrightarrow{AB}$ meets $\overrightarrow{QS}$. Hence, $\overrightarrow{AR} \parallel \overrightarrow{QS}$.

On the other hand, if A lies on the extension of $\overrightarrow{RP}$ through P, and if $\overrightarrow{AB}$ is a ray interior to ∢RAQ, then $\overrightarrow{AB}$ meets $\overrightarrow{PQ}$ in a point B (Theorem III-17). Let $\overrightarrow{PC}$ be parallel to $\overrightarrow{AB}$. $\overrightarrow{PC}$ meets $\overrightarrow{QS}$ in some point C. Apply Pasch's Axiom to △PQC to show that $\overrightarrow{AB}$ meets $\overrightarrow{QC}$.

The reader may quite easily extend Lemma V-3 to obtain the following.

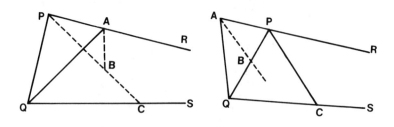

Figure V - 5.

ANGLE OF PARALLELISM

Theorem V-4
If (rays) $\overrightarrow{PR} \parallel \overrightarrow{QS}$, then $\overrightarrow{AR} \parallel \overrightarrow{BS}$ for any A on $\overleftrightarrow{PR}$ and B on $\overleftrightarrow{QS}$, provided R is not between P and A and S is not between Q and B.

Exercises V-2
1. Let $\mathscr{X}$ and $\mathscr{Y}$ be the sets defined in Section 1. Prove that each angle of $\mathscr{X}$ is smaller than every angle of $\mathscr{Y}$.
2. Prove Theorem V-4.

3. PARALLELISM OF LINES—THE ANGLE CRITERION

Inasmuch as parallelism of rays is independent of the initial points, we may speak of the parallelism of two lines, in a specified direction.

Definition V-5
Line $\overleftrightarrow{BB'}$ is parallel to line $\overleftrightarrow{AA'}$ (in the direction $\overrightarrow{AA'}$) if the following two conditions are met:

a. $\overleftrightarrow{BB'}$ does not meet $\overleftrightarrow{AA'}$
b. for some P on $\overleftrightarrow{BB'}$ and Q on $\overleftrightarrow{AA'}$, any ray $\overrightarrow{PS}$ interior to $\sphericalangle B'PQ$ (opening in the direction of parallelism) will meet $\overrightarrow{QA'}$

Condition (b) is often called the *angle criterion*. By Theorem V-4, if (b) holds for one choice of P and Q, it holds for all choices. Note that the angle alluded to in the angle criterion has its vertex on $\overleftrightarrow{BB'}$, the *first* line. Hence it is not clear that $\overleftrightarrow{BB'} \parallel \overleftrightarrow{AA'}$ and $\overleftrightarrow{AA'} \parallel \overleftrightarrow{BB'}$ mean the same thing.

We shall usually omit the phrase "in the direction $\overrightarrow{AA'}$" and let the ordering of the symbols—$\overleftrightarrow{AA'}$ or $\overleftrightarrow{A'A}$, as the case may be—indicate the direction for us. As Theorems V-7 and V-8 will assert, parallelism in a given direction is a symmetric and transitive relation.

According to the remarks preceding Lemma V-3, in Lobachevskian geometry, through a point P not on a line $\overleftrightarrow{AA'}$ there are exactly two lines $\overleftrightarrow{BB'}$ and $\overleftrightarrow{CC'}$ parallel respectively to $\overleftrightarrow{AA'}$ and $\overleftrightarrow{A'A}$. These parallels are characterized by the properties (Fig. V-6)

1. any line through P which passes through the interior of $\sphericalangle C'PB'$ (containing the perpendicular $\overline{PQ}$) meets $\overleftrightarrow{AA'}$, while
2. any line through P which passes through the interiors of $\sphericalangle BPC'$ and $\sphericalangle CPB'$ does not meet $\overleftrightarrow{AA'}$

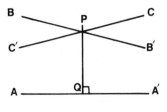

Figure V - 6.

The (infinitely many) lines of this second class are said to be *nonintersecting* or *divergent* with respect to $\overleftrightarrow{AA'}$.

Remark: Note that through a given point not on a given line there is only one line parallel to the given line in *each* of the two directions.

Exercises V-3

1. Sketch examples of parallel and divergent lines in

   a. the Klein-Beltrami Model
   b. the Poincaré Disk Model

(These models were described in Section IV-2.) In each case, what is the interpretation of the word "parallel" in the sense of Definition V-5? Convince yourself that the angle criterion is satisfied.

2. Recall that "lines" in the Poincaré Upper Half-Plane Model (Section IV-2) are open semi-circles centered on the x axis or open rays perpendicular to the x axis. $\overleftrightarrow{PQ}$ in Figure IV-9 is a "line" of the latter type.

   a. Describe (as e.g., with a sketch) all "lines" parallel to $\overleftrightarrow{PQ}$, in the sense of Definition V-5
   b. Describe all "lines" parallel to $\overleftrightarrow{QP}$
   c. Describe all "lines" which are divergent with $\overleftrightarrow{PQ}$

4. BISECTOR OF THE STRIP

If $\overleftrightarrow{AA'} \parallel \overleftrightarrow{BB'}$, we shall prove that there exists a unique line $\overleftrightarrow{CC'}$ between them and relative to which the lines are symmetrically located (in the sense made explicit in Theorem V-6). We shall use this line, called the *bisector of the strip* between $\overleftrightarrow{AA'}$ and $\overleftrightarrow{BB'}$, to show that $\overleftrightarrow{AA'} \parallel \overleftrightarrow{BB'}$ implies $\overleftrightarrow{BB'} \parallel \overleftrightarrow{AA'}$ (i.e., parallelism is a symmetric relation). The bisector of the strip will be useful later as well.

BISECTOR OF THE STRIP

Theorem V-6

If $\overleftrightarrow{AA'} \parallel \overleftrightarrow{BB'}$, there exists a unique line $\overleftrightarrow{CC'}$ with the following properties (see Figure V-7):

a. $\overleftrightarrow{AA'}$ and $\overleftrightarrow{BB'}$ lie on opposite sides of $\overleftrightarrow{CC'}$, and $\overleftrightarrow{AA'} \parallel \overleftrightarrow{CC'}$
b. Each point M on $\overleftrightarrow{CC'}$ is equidistant from $\overleftrightarrow{AA'}$ and $\overleftrightarrow{BB'}$, and the angle formed at M by the perpendiculars from M to $\overleftrightarrow{AA'}$ and $\overleftrightarrow{BB'}$ is bisected by $\overleftrightarrow{CC'}$
c. For each point P on $\overleftrightarrow{AA'}$, there exists a unique point Q on $\overleftrightarrow{BB'}$ such that $\overleftrightarrow{CC'}$ is the perpendicular bisector of $\overline{PQ}$. We write $P \triangleq Q$ (a notation used by J. Bolyai)
d. If $P \triangleq Q$, then $\overleftrightarrow{AA'}$ and $\overleftrightarrow{BB'}$ make congruent acute angles with $\overline{PQ}$, on the side of $\overline{PQ}$ which is in the direction of parallelism
e. If $P \triangleq Q$ and $P' \triangleq Q'$, then $PP' = QQ'$ (P,P' on $\overleftrightarrow{AA'}$; Q,Q' on $\overleftrightarrow{BB'}$)

Proof: Join any point D of $\overleftrightarrow{AA'}$ to any point E of $\overleftrightarrow{BB'}$. Draw the bisectors of ∢A'DE and ∢B'ED. The former meets $\overrightarrow{EB'}$ since $\overleftrightarrow{AA'} \parallel \overleftrightarrow{BB'}$. Therefore, by Theorem III-17, the two angle bisectors meet in a point O. From O drop perpendiculars $\overline{OF}$ to $\overleftrightarrow{AA'}$, $\overline{OG}$ to $\overleftrightarrow{BB'}$, and $\overline{OH}$ to $\overline{DE}$. By SAA, $\triangle ODF \cong \triangle ODH$ and $\triangle OEH \cong \triangle OEG$, and so OF = OH = OG. Draw through O the bisector $\overleftrightarrow{CC'}$ of ∢FOG, as in Figure V-7. We will now show that $\overleftrightarrow{CC'}$ is the desired bisector of the strip between $\overleftrightarrow{AA'}$ and $\overleftrightarrow{BB'}$.

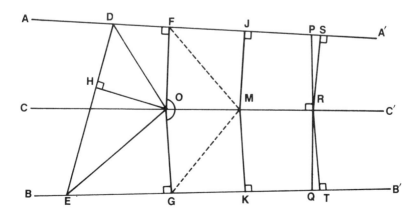

Figure V - 7.

First, $\overleftrightarrow{CC}'$ does not meet $\overleftrightarrow{AA}'$ or $\overleftrightarrow{BB}'$, for if $\overleftrightarrow{CC}'$ intersected either of these lines in a point N, then the congruence of $\triangle$FON and $\triangle$GON would imply that $\overleftrightarrow{FN}$ and $\overleftrightarrow{GN}$ coincide with $\overleftrightarrow{AA}'$ and $\overleftrightarrow{BB}'$, respectively, which is impossible because $\overleftrightarrow{AA}'$ and $\overleftrightarrow{BB}'$ are parallel. Therefore, since F and G lie on opposite sides of $\overleftrightarrow{CC}'$ (because $\overline{FG}$ crosses $\overleftrightarrow{CC}'$), the same is true of $\overleftrightarrow{AA}'$ and $\overleftrightarrow{BB}'$. Moreover the angle criterion is satisfied since $\overrightarrow{DE}$ meets $\overleftrightarrow{CC}'$ and any ray through D interior to ⊀A'DE meets $\overleftrightarrow{BB}'$ and hence $\overleftrightarrow{CC}'$. Thus $\overleftrightarrow{AA}' \parallel \overleftrightarrow{CC}'$ and (a) is proved.

Next, we show that any point M on $\overleftrightarrow{CC}'$ is equisdistant from $\overleftrightarrow{AA}'$ and $\overleftrightarrow{BB}'$. To see this, drop perpendiculars $\overline{MJ}$ to $\overleftrightarrow{AA}'$ and $\overline{MK}$ to $\overleftrightarrow{BB}'$ and draw $\overline{FM}$ and $\overline{GM}$. By SAS, $\triangle$FOM $\cong \triangle$GOM, whence FM = GM, and ⊀MFJ $\cong$ ⊀MGK. Therefore, $\triangle$FJM $\cong \triangle$GKM, and so JM = MK and ⊀JMO $\cong$ ⊀KMO. (b) is proved.

Now let P be any point of $\overleftrightarrow{AA}'$. Draw $\overline{PR} \perp \overleftrightarrow{CC}'$, $\overline{RS} \perp \overleftrightarrow{AA}'$, and $\overline{RT} \perp \overleftrightarrow{BB}'$. Since ⊀A'PR is acute (it is Π(PR)), S is on ray $\overrightarrow{PA}'$. On ray $\overrightarrow{TB}$, mark Q so that TQ = PS and draw $\overline{RQ}$. Then $\triangle$PRS $\cong \triangle$QRT, ⊀PRS $\cong$ ⊀TRQ, and also ⊀SRC' $\cong$ ⊀TRC' (by (b)). Accordingly, ⊀QRC' $\cong$ ⊀PRC' = $\pi/2$, so that PRQ is a line. Since ⊀RPS $\cong$ ⊀RQT and PR = RQ, we have established (c) and (d).

The proof of (e) and the uniqueness of $\overleftrightarrow{CC}'$ are left for the exercises.

It is now a simple matter to prove the symmetry of parallelism.

Theorem V-7

If $\overleftrightarrow{AA}' \parallel \overleftrightarrow{BB}'$, then $\overleftrightarrow{BB}' \parallel \overleftrightarrow{AA}'$.

Proof. Let P $\stackrel{\simeq}{=}$ Q, where P is on $\overleftrightarrow{AA}'$ and Q is on $\overleftrightarrow{BB}'$ (Fig. V-8). It suffices to show that any ray $\overrightarrow{QR}$ interior to ⊀B'QP meets $\overrightarrow{PA}'$. To this end, draw ray $\overrightarrow{PT}$ interior to ⊀A'PQ so that ⊀QPT $\cong$ ⊀PQR. By

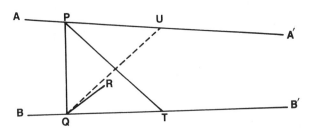

Figure V - 8.

BISECTOR OF THE STRIP 175

the angle criterion, $\overrightarrow{PT}$ meets $\overrightarrow{QB'}$ in a point T. Let U ≜ T and draw $\overline{QU}$. The congruence of $\triangle PQT$ and $\triangle QPU$ follows from (d) and (e) of the preceding Theorem and SAS. But then ∡PQU ≅ ∡QPT ≅ ∡PQR, so that $\overleftrightarrow{QU}$ and $\overleftrightarrow{QR}$ are the same line. $\overleftrightarrow{QR}$ therefore meets $\overrightarrow{PA'}$.

Theorem V-8
Let $\overleftrightarrow{AA'}$, $\overleftrightarrow{BB'}$, and $\overleftrightarrow{CC'}$ be distinct lines. Then if $\overleftrightarrow{AA'}$ ∥ $\overleftrightarrow{BB'}$ and $\overleftrightarrow{BB'}$ ∥ $\overleftrightarrow{CC'}$, then $\overleftrightarrow{AA'}$ ∥ $\overleftrightarrow{CC'}$.

Proof. There are two cases to consider: (1) $\overleftrightarrow{AA'}$ and $\overleftrightarrow{CC'}$ on opposite sides of $\overleftrightarrow{BB'}$; and (2) $\overleftrightarrow{AA'}$ and $\overleftrightarrow{CC'}$ on the same side of $\overleftrightarrow{BB'}$.

Case 1: Let P lie on $\overleftrightarrow{AA'}$, R lie on $\overleftrightarrow{CC'}$, and let Q be the intersection of $\overleftrightarrow{PR}$ with $\overleftrightarrow{BB'}$. Since $\overleftrightarrow{AA'}$ ∥ $\overleftrightarrow{BB'}$, any ray $\overrightarrow{PT}$ interior to ∡A'PQ will meet $\overleftrightarrow{BB'}$ in a point T (Fig. V-9). Draw $\overline{RT}$. $\overrightarrow{PT}$ extended through T enters ∡B'TR and therefore meets $\overleftrightarrow{CC'}$ since $\overleftrightarrow{BB'}$ ∥ $\overleftrightarrow{CC'}$. Accordingly, $\overleftrightarrow{AA'}$ ∥ $\overleftrightarrow{CC'}$.

Case 2: $\overleftrightarrow{AA'}$ and $\overleftrightarrow{CC'}$ cannot meet, since otherwise we would have two parallels to $\overleftrightarrow{BB'}$ in the same direction through their point of intersection. We may therefore assume, without loss of generality, that $\overleftrightarrow{CC'}$ is between $\overleftrightarrow{AA'}$ and $\overleftrightarrow{BB'}$.* We can then prove the stronger assertion that $\overleftrightarrow{AA'}$ and $\overleftrightarrow{BB'}$ are on opposite sides of $\overleftrightarrow{CC'}$. Choose arbitrary points M, N, and P on $\overleftrightarrow{AA'}$, $\overleftrightarrow{BB'}$, and $\overleftrightarrow{CC'}$, respectively (Fig. V-10). Of the two angles ∡B'NP and ∡B'NM which open in the direction of parallelism, choose the smaller and draw ray $\overrightarrow{NQ}$ in its

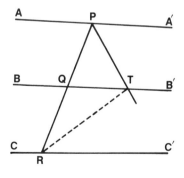

Figure V - 9.

---

*This is because, by Theorem V-7, the statement of Theorem V-8 is symmetric in $\overleftrightarrow{AA'}$ and $\overleftrightarrow{CC'}$.

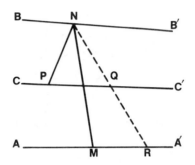

Figure V - 10.

interior. Since $\overleftrightarrow{BB'} \parallel \overleftrightarrow{CC'}$, and $\overleftrightarrow{BB'} \parallel \overleftrightarrow{AA'}$, $\overrightarrow{NQ}$ will meet $\overleftrightarrow{CC'}$ in Q and $\overleftrightarrow{AA'}$ in a point R. Thus, N and R are on opposite side of $\overleftrightarrow{CC'}$. The same must be true of $\overleftrightarrow{AA'}$ and $\overleftrightarrow{BB'}$ since neither of these lines cross $\overleftrightarrow{CC'}$.

Now to show that $\overleftrightarrow{AA'} \parallel \overleftrightarrow{CC'}$, observe that $\overleftrightarrow{MN}$ intersects $\overleftrightarrow{CC'}$ and so is a transversal to $\overleftrightarrow{AA'}$ and $\overleftrightarrow{CC'}$. Any ray $\overrightarrow{MS}$ interior to ∢NMA' will meet $\overleftrightarrow{BB'}$ and hence must cross $\overleftrightarrow{CC'}$. This establishes the angle criterion and the parallelism of $\overleftrightarrow{AA'}$ and $\overleftrightarrow{CC'}$.

Note that as a by-product we have shown that if two lines are parallel to a third line in a given direction, then two of these three lines lie on opposite sides of the third.

Exercises V-4

    1. Prove part (e) of Theorem V-6.

    2. Prove that the bisector of the strip between two parallel lines is unique. (Hint: in Fig. V-7, let P be a point of $\overleftrightarrow{AA'}$. Let P ≐ Q relative to one bisector of the strip, and let P ≐ Q' relative to another. Q may or may not coincide with Q'.)

5. PROPERTIES OF Π(x)

We proceed now to derive some further properties of parallels and of the angle of parallelism Π(x), for which we shall eventually derive an explicit formula in terms of the distance x from the point to the line. Many of the following theorems will seem to contradict what seem to be obvious features of the diagrams. This stems from the fact that ordinary points and ordinary lines of a flat (Euclidean) plane cannot serve as a model for the postulates of Lobachevskian geometry. Since a proper model necessarily involves a curved space, the lines in our

PROPERTIES OF $\Pi(x)$

diagrams will often be drawn curved to conform better to the non-Euclidean facts. In any event, diagrams are merely heuristic or mnemonic tools, and one should not utilize any features of them that have not been deduced (or are not clearly deducible) from the postulates.

The strangeness of the following theorem will perhaps be diminished if you note that every angle of parallelism provides an illustration.

Theorem V-9

Given an acute angle, $\sphericalangle POQ$, there exists a unique line perpendicular to side $\overrightarrow{OP}$ and parallel to side $\overrightarrow{OQ}$.

Proof. Suppose that every perpendicular erected on a point of $\overrightarrow{OP}$ intersected $\overrightarrow{OQ}$. In Figure V-11, choose points A_0, A_1, A_2, A_3, ... on $\overrightarrow{OP}$ so that $OA_0 = A_0 A_1$, $OA_1 = A_1 A_2$, $OA_2 = A_2 A_3$, etc., or equivalently, $OA_n = 2^n OA_0$. Let the perpendiculars to $\overrightarrow{OP}$ erected at the points in this sequence meet $\overrightarrow{OQ}$ in the points B_0, B_1, B_2, B_3, etc. Let δ denote the defect of $\triangle OA_0 B_0$. Then, since defect is additive, and since for each $n \geq 0$, $\triangle OA_n B_n \cong \triangle A_{n+1} A_n B_n$, we have

$\mathscr{D}(OA_1 B_1) > 2\delta$
$\mathscr{D}(OA_2 B_2) > 2\mathscr{D}(OA_1 B_1) > 2^2 \delta$
$\mathscr{D}(OA_3 B_3) > 2\mathscr{D}(OA_2 B_2) > 2^3 \delta$

.
.
.

$\mathscr{D}(OA_n B_n) > 2\mathscr{D}(OA_{n-1} B_{n-1}) > 2^n \delta.$

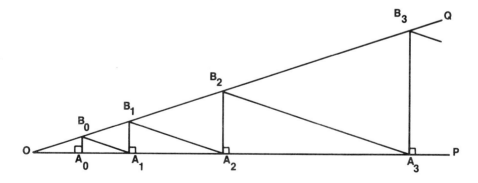

Figure V-11.

For n large enough, $2^n \delta$ will exceed π, since $\delta > 0$. This is impossible, since $\mathscr{D}(OA_n B_n) = \pi - \mathscr{S}(OA_n B_n) < \pi$, for all n. Hence, in Lobachevskian geometry, there must exist at least one point on $\overrightarrow{OP}$ at which the erected perpendicular will fail to meet $\overrightarrow{OQ}$.

Now divide the points of ray $\overrightarrow{OP}$ into two disjoint classes. Let $\mathscr{X} = \{$ X on $\overrightarrow{OP}$ such that the perpendicular to $\overrightarrow{OP}$ at X meets $\overrightarrow{OQ} \}$ and let $\mathscr{Y} = \{$ Y on $\overrightarrow{OP}$ such that the perpendicular to $\overrightarrow{OP}$ at Y does not meet $\overrightarrow{OQ}\}$ (Fig. V-12). $\mathscr{X}$ is not empty since we may always drop a perpendicular to $\overrightarrow{OP}$ from any point of $\overrightarrow{OQ}$. We saw above that $\mathscr{Y}$ is not empty. Also, by Pasch's axiom, every point of $\mathscr{X}$ is between O and any point of $\mathscr{Y}$ (check this).

By Dedekind's postulate, there is on $\overrightarrow{OP}$ a unique point C which forms the boundary between $\mathscr{X}$ and $\mathscr{Y}$. Moreover, C belongs to $\mathscr{Y}$, for if the perpendicular $\overleftrightarrow{CD}$ were to meet $\overrightarrow{OQ}$ in a point D, then by dropping a perpendicular to $\overrightarrow{OP}$ from a point on ray $\overrightarrow{DQ}$, we would obtain a point of $\mathscr{X}$ that lies on ray $\overleftrightarrow{CP}$, a contradiction.

To show that $\overleftrightarrow{CD} \parallel \overrightarrow{OQ}$, let $\overrightarrow{CE}$ be a ray in the interior of $\angle OCD$, where E is within figure QOCD. Drop $\overline{EF} \perp \overline{OC}$. Since F is in $\mathscr{X}$, $\overline{FE}$ extended through E meets $\overrightarrow{OQ}$ in a point G. Then, by Pasch's axiom applied to $\triangle OFG$, ray $\overrightarrow{CE}$ must meet $\overrightarrow{OQ}$, and so the angle criterion is satisfied.

It is clear that $\overleftrightarrow{CD}$ is the only perpendicular to $\overrightarrow{OP}$ which is parallel to $\overrightarrow{OQ}$ since any other would necessarily be parallel to $\overleftrightarrow{CD}$

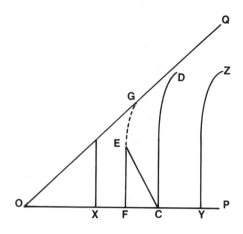

Figure V - 12.

PROPERTIES OF Π(x)

by Theorem V-8, and lines with a common perpendicular cannot be parallel in Lobachevskian geometry [Π(x) is always acute].

Since, in the preceding, $\angle POQ = \Pi(OC)$, we have the following.

Corollary V-10

Every acute angle is an angle of parallelism. The function Π maps the positive real numbers *onto* the interval $(0, \pi/2)$.

By bisecting any angle between 0 and π and using Corollary V-10, the reader may prove the following (see Exercise 2).

Corollary V-11

Given any (non-straight) angle, there exists a unique line parallel to both of its sides.

Theorem V-12

The perpendicular distance from points on one of two parallel lines to the other decreases and approaches zero in the direction of parallelism and increases without bound in the opposite direction.

Proof. In the birectangular quadrilateral in Figure V-13, $\overleftrightarrow{AB} \parallel \overleftrightarrow{CD}$. Hence $\angle A < \pi/2 < \angle B$. It then follows from absolute geometry (Exercise 2 of Section IV-4) that the side $\overline{AC}$ opposite the larger angle, $\angle B$, is greater than the side $\overline{BD}$ opposite the smaller angle, $\angle A$. Hence, the distance from one of two parallel lines to the other is strictly decreasing in the direction of parallelism. However, we wish to prove the stronger result that for any given distance x, no matter how large or how small, there is a point on either one of two parallel lines whose distance from the other equals x.

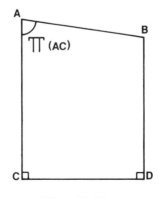

Figure V - 13.

To the end, let $\overleftrightarrow{AA'}$ and $\overleftrightarrow{BB'}$ be two parallel lines and from an arbitrary point P on $\overleftrightarrow{AA'}$ drop $\overline{PQ} \perp \overleftrightarrow{BB'}$. If PQ = x, we are done. Otherwise, on $\overrightarrow{QP}$ (Fig. V-14a), or $\overrightarrow{QP}$ extended through P (Fig. V-14b) mark off QR = x, and draw $\overrightarrow{RS} \parallel \overrightarrow{B'B}$ (opposite to the direction of parallelism of $\overleftrightarrow{AA'}$ and $\overleftrightarrow{BB'}$). Then $\overrightarrow{RS}$, or $\overline{SR}$ extended through R, will meet $\overleftrightarrow{AA'}$ in a point T. This may be seen in the case PQ > QR by drawing $\overrightarrow{RR'} \parallel \overrightarrow{BB'}$ and applying the angle criterion to ∡PRR'. In the case PQ < QR, draw $\overrightarrow{PP'} \parallel \overrightarrow{B'B}$ and apply the angle criterion to ∡RPP'.

Draw $\overline{TU} \perp \overleftrightarrow{BB'}$ and mark off V on $\overleftrightarrow{AA'}$ so that T lies between P and V, and so that TV = RT. Drop $\overline{VW} \perp \overleftrightarrow{BB'}$. Since ∡STU ≅ ∡A'TU = Π(TU), it is easily checked (draw $\overline{RU}$ and $\overline{VU}$ and use congruent triangles) that VW = RQ = x, as desired.

Exercises V-5

1. Show that if Π(x) = α, a constant, then α must be a right angle, and so Euclid's postulate holds.

2. Prove Corollary V-11. (If $\overrightarrow{OR}$ is the bisector of ∡POQ, then Theorem V-9 guarantees the existence of a line perpendicular to $\overrightarrow{OR}$ and parallel to $\overrightarrow{OP}$, and of a line perpendicular to $\overrightarrow{OR}$ and parallel to $\overrightarrow{OQ}$. You must show these lines coincide.)

3. (a) Let ∡POQ be any given (non-straight) angle. Prove that in Lobachevskian geometry there exists a point X in the angle's interior with the property that no line through X (other than $\overleftrightarrow{OX}$) intersects both $\overrightarrow{OQ}$ and $\overrightarrow{OP}$. (b) Show that no such angle and point exist in Euclidean geometry. (c) What was Legendre's "unwarranted assumption," cited in connection with Figure IV-15?

4. Show that the perpendicular distances from points on one side of an acute angle to the other side increase without bound as

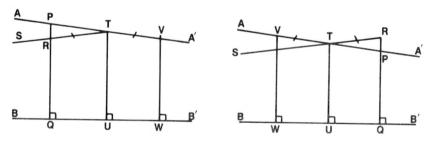

Figure V - 14.

PROPERTIES OF Π(x)

these points recede from the vertex. (Hint: In the figure on this page, where $\overleftrightarrow{CD} \parallel \overrightarrow{OQ}$, we may mark off on $\overrightarrow{CD}$ a segment $\overline{CE}$ longer than any preassigned length. Show that the perpendicular to $\overleftrightarrow{CD}$ at E must meet $\overrightarrow{OQ}$. Drop $\overline{FG}$. Use Exercise 2 of Section IV-4.)

5. Suppose $\overleftrightarrow{AA'} \parallel \overleftrightarrow{BB'}$. Show that there exists a line perpendicular to $\overleftrightarrow{BB'}$ and parallel to $\overleftrightarrow{A'A}$.

6. Let ℓ and m be parallel lines. Prove that the feet of all the perpendiculars dropped to m from all points of ℓ comprise an open ray of m.

6. IDEAL POINTS

In view of the preceding theorem, one is tempted to think of parallel lines as converging to a "point at infinity," or "meeting at infinity." Of course, there is no such point as infinity, and two parallel lines, by the very definition of parallelism, have no intersection. Nevertheless, we can indulge our imagination somewhat by introducing the following terminology.

Definition V-13

The family consisting of a line and all lines parallel to it in a given direction is called an *ideal point*.

An ideal point Ω is obviously determined by any one line belonging to the family Ω, and by the direction of parallelism. Through any point of the plane there passes a unique member of Ω. Thus, an ideal point is not a point at all, but a set of lines. Nevertheless, we shall use jargon that will be suggestive of a point. If a line ℓ is a member of an ideal point Ω, we shall say that "ℓ contains Ω," "ℓ passes through Ω," or "Ω is on ℓ." Thus to say that two lines "meet" in an ideal point Ω would mean that they both belong to the same family Ω of parallel lines. Parallel lines "meet" in an ideal point. A (real) point A and an ideal point Ω determine a unique line, which we shall

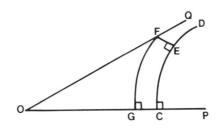

denote $\overleftrightarrow{A\Omega}$. Every line contains two ideal points (since every line determines two directions of parallelism).

The reader should guard against taking this terminology literally and should retain the thought that an ideal point is, strictly speaking, a family of lines, all parallel in the same direction.

Exercises V-6
1. Describe the interpretation of the term *ideal point* in

   a. the Klein-Beltrami Model
   b. the Poincaré Disk Model
   c. the Poincaré Upper Half-Plane Model

all described in Section IV-2; cf. Exercises 1 and 2 of Section 3.

2. How would you define *ideal point* in Euclidean geometry? How many ideal points does a line contain in Euclidean geometry?

7. IDEAL TRIANGLES

In Lobachevskian geometry, the figure formed by two parallel rays $\overrightarrow{A\Omega}$ and $\overrightarrow{B\Omega}$, and a transversal $\overline{AB}$ may be viewed as a triangle with an ideal point for one of its vertices. Such "ideal triangles" have many of the properties of ordinary triangles. In this section, Ω, Ω', etc. are ideal points and all other points are real.

Theorem V-14

If a line enters the interior of the ideal triangle $\triangle AB\Omega$ through one of its three vertices, then it will intersect the opposite side.

Proof. If the line enters the triangle through A or B, the conclusion is immediate from the angle criterion. If the line enters through (i.e., belongs to the family) Ω, let P be a point on the line within the triangle (Fig. V-15). By the angle criterion, $\overrightarrow{AP}$ meets $\overrightarrow{B\Omega}$ in a point C, and we may now apply Pasch's axiom to $\triangle ABC$, to deduce that the line intersects $\overline{AB}$.

An easily established consequence of Theorem V-14 is this generalization of Pasch's axiom to ideal triangles. The proof is left as an exercise for the reader.

Corollary V-15

If a line intersects one of the sides of an ideal triangle but does not pass through any one of the three vertices, then it intersects one of the other two sides.

IDEAL TRIANGLES

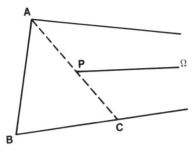

Figure V - 15.

Theorem V-16

If, in ideal triangle $\triangle AB\Omega$, side $\overline{AB}$ is produced through B to a point C, then the exterior angle, $\measuredangle CB\Omega$, is greater than the opposite interior angle, $\measuredangle BA\Omega$.

Proof. Since $\measuredangle AB\Omega + \measuredangle BA\Omega \leq \pi$ (cf. the paragraph following Definition V-1), the exterior angle $\measuredangle CB\Omega$ (= $\pi - \measuredangle AB\Omega$) is not less than $\measuredangle BA\Omega$ (Fig. V-16). If $\measuredangle CB\Omega \cong \measuredangle BA\Omega$, then from the midpoint M of $\overline{AB}$ drop $\overline{MP} \perp \overleftrightarrow{A\Omega}$ and $\overline{MQ} \perp \overleftrightarrow{B\Omega}$. Then $\measuredangle PAM \cong \measuredangle MBQ$ and so $\triangle PAM \cong \triangle QBM$ by SAA. Consequently, $\measuredangle PMA \cong \measuredangle BMQ$, so that PMQ is a line and so a common perpendicular to $\overleftrightarrow{A\Omega}$ and $\overleftrightarrow{B\Omega}$. This contradicts their parallelism. Hence we are left with the alternative that $\measuredangle CB\Omega > \measuredangle BA\Omega$.

If we pretend that there is an angle of zero magnitude at the ideal vertex (Exercise 7), the following generalizes the SAA congruence theorem to ideal triangles.

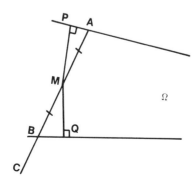

Figure V - 16.

Theorem V-17

In ideal triangles $\triangle AB\Omega$ and $\triangle A'B'\Omega'$, let $AB = A'B'$ and $\angle AB\Omega \cong \angle A'B'\Omega'$. Then $\angle BA\Omega \cong \angle B'A'\Omega'$.

Proof. Suppose, on the contrary, that $\angle BA\Omega \not\cong \angle B'A'\Omega'$. Let's say, for definiteness, that $\angle BA\Omega > \angle B'A'\Omega'$. Then in the interior of $\angle BA\Omega$ draw ray $\overrightarrow{AC}$ such that $\angle BAC \cong \angle B'A'\Omega'$ (Fig. V-17). By the angle criterion, $\overrightarrow{AC}$ meets $B\Omega$ in C. On $\overrightarrow{B'\Omega'}$ mark off $B'C' = BC$. The congruence of $\triangle ABC$ and $\triangle A'B'C'$ yields the contradiction of $\angle B'A'\Omega' \cong \angle BAC \cong \angle B'A'C' < \angle B'A'\Omega'$. Consequently, $\angle BA\Omega$ must be congruent to $\angle B'A'\Omega'$.

Exercises V-7

1. Prove Corollary V-15.
2. Show that in Lobachevskian geometry, when two parallel lines are cut by a transversal then the sum of the interior angles on the side of the transversal in the direction of parallelism is less than two right angles.

(The following exercises are to be taken in the context of Lobachevskian geometry.)

3. (AAA Theorem for Ideal Triangles) In ideal triangles $\triangle AB\Omega$ and $\triangle A'B'\Omega'$, assume $\angle A \cong \angle A'$, $\angle B \cong \angle B'$. Prove $AB = A'B'$.
4. Prove: if, in ideal triangles $\triangle AB\Omega$ and $\triangle A'B'\Omega'$, $\angle A \cong \angle A'$ and $AB > A'B'$, then $\angle B < \angle B'$.
5. (a) Prove Pasch's Axiom for a triangle with two ideal vertices. (b) Prove it for a triangle with three ideal vertices.
6. Prove: if, in ideal triangles $\triangle AB\Omega$ and $\triangle A'B'\Omega'$, $\angle A \cong \angle B$, $\angle A' \cong \angle B'$, and $AB = A'B'$, then $\angle A \cong \angle A'$.
7. Let C be a variable point on side $B\Omega$ of ideal triangle $\triangle AB\Omega$. Show that as C recedes farther and farther away from B in the direc-

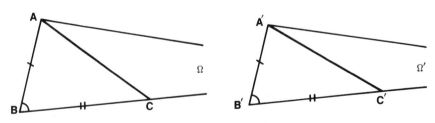

Figure V - 17.

IDEAL TRIANGLES

tion of Ω, $\angle ACB$ becomes arbitrarily small, and, if $\angle A \cong \angle B$, then $\angle BAC \to \Pi(AB/2)$ (cf. Exercise 12 of Section IV-2).

8. Given ideal triangle $\triangle AB\Omega$ and parallel lines $\overleftrightarrow{P\Omega}'$ and $\overleftrightarrow{Q\Omega}'$, show that there exist points A' on $\overrightarrow{P\Omega}'$ and B' on $\overrightarrow{Q\Omega}'$ such that $\triangle AB\Omega \cong \triangle A'B'\Omega'$. (Hint: Theorem V-12).

9. In Euclidean geometry, we may define an *ideal point* to be the family consisting of a line and all lines parallel to it. Two rays $\overrightarrow{AR}$ and $\overrightarrow{BS}$ will be called *parallel* if they are contained in parallel lines and if R and S lie in the same half-plane of $\overleftrightarrow{AB}$. Ideal triangles are defined as in this section. With these definitions, which of the results of this section are valid in Euclidean geometry and which are false?

8. MORE PROPERTIES OF $\Pi(x)$

The theorems of the preceding section provide us with two fundamental facts concerning the angle of parallelism.

Theorem V-18
    The angle of parallelism depends only upon the distance from the point to the line (and not where they are situated in the plane).
    Proof. In Figure V-18, PQ = P'Q' and the angles at Q and Q' are right. Hence $\triangle PQ\Omega \cong \triangle P'Q'\Omega'$ (Theorem V-17), and so $\Pi(PQ) \cong \Pi(P'Q')$.

Theorem V-19
    $\Pi(x)$ is a strictly decreasing function of x, the distance from the point to the line.
    Proof. Let $\overline{PQ}$ be perpendicular to $\overrightarrow{Q\Omega}$ and let P' be a point of segment $\overline{PQ}$. Draw $\overrightarrow{P\Omega}$ and $\overrightarrow{P'\Omega}$ (Fig. V-19). Then P'Q < PQ, and

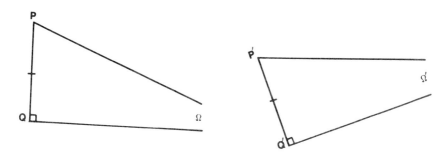

Figure V - 18.

186 FUNDAMENTALS OF LOBACHEVSKIAN GEOMETRY

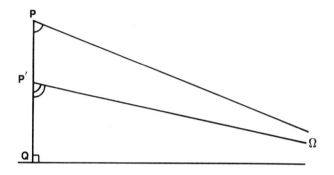

Figure V - 19.

since $\angle QP'\Omega = \Pi(P'Q)$ is an exterior angle of $\triangle PP'\Omega$, we have $\Pi(P'Q) > \Pi(PQ)$, by Theorem V-16.

9. DIVERGENT LINES

Recall that lines which are neither intersecting nor parallel are called *divergent*.

Theorem V-20

Two divergent lines have a unique common perpendicular.
Proof. Let $\overleftrightarrow{AA'}$ and $\overleftrightarrow{\Omega\Omega'}$ be two divergent lines, as in Figure V-20. From an arbitrary point B on $\overleftrightarrow{AA'}$, draw the parallels $\overrightarrow{B\Omega}$ and $\overrightarrow{B\Omega'}$. By Theorem V-9, we may erect on $\overleftrightarrow{AA'}$ perpendiculars $\overrightarrow{C\Omega}$ and $\overrightarrow{C'\Omega'}$ which are parallel to $\overrightarrow{B\Omega}$ and $\overrightarrow{B'\Omega'}$, respectively. From the midpoint P of $\overline{CC'}$ draw $\overline{PQ} \perp \overleftrightarrow{\Omega\Omega'}$. We now show $\overline{PQ} \perp \overleftrightarrow{AA'}$ as well as $\perp \overleftrightarrow{\Omega\Omega'}$.

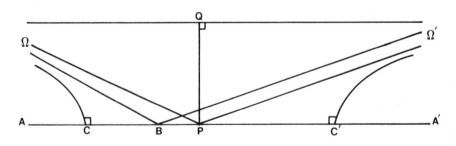

Figure V - 20.

DIVERGENT LINES

To this end draw $\overrightarrow{P\Omega}$ and $\overrightarrow{P\Omega'}$. Then $\angle \Omega PQ \cong \angle \Omega'PQ = \Pi(PQ)$, and $\angle \Omega PC = \Pi(PC) = \Pi(PC') = \angle \Omega'PC'$. It follows that $\angle QPA' \cong \angle QPA = \pi/2$, and so $\overline{PQ}$ is a common perpendicular to $\overleftrightarrow{AA'}$ and $\overleftrightarrow{\Omega\Omega'}$. The impossibility of the existence of a rectangle implies the uniqueness of the common perpendicular.

Theorem V-21

The perpendicular distance from points on either one of two divergent lines to the other increases without bound as these points recede from the common perpendicular on either side.

Proof. Suppose $\overline{MN}$ is the common perpendicular to two divergent lines m and n, as in Figure V-21, with M on m and N on n.

You can readily verify that the distance between m and n increases with increasing distance from $\overline{MN}$ (compare the proof of Theorem V-12). We proceed now to show that the distance increases *without bound*.

Let Ω be either one of the two ideal points on m and draw $\overrightarrow{N\Omega}$. By Exercise 4 of Section V-5, we may choose R on N such that the perpendicular distance RQ from R to n is arbitrarily large. $\overline{QR}$ extended through R lies within $\angle MR\Omega$ and so, by the angle criterion, $\overleftrightarrow{QR}$ meets m at P. Thus PQ > RQ, which was chosen arbitrarily large.

Exercises V-9

1. Show that if two lines are cut by a transversal so that alternate interior angles are congruent, then the lines are divergent. (Hint: see the proof of Theorem V-16.)

2. Let ℓ and m be divergent lines with common perpendicular $\overline{AB}$, where A is on ℓ and B is on m. Prove that any transvergal to ℓ

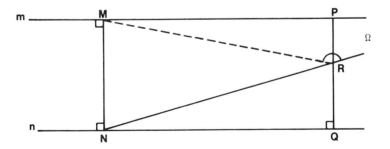

Figure V - 21.

and m which passes through the midpoint of $\overline{AB}$ makes with ℓ and m congruent alternate interior angles.

3. Show that if opposite sides of a quadrilateral are congruent, then opposite sides are divergent.

4. Let ℓ and m be divergent lines. Show that the feet of all perpendiculars dropped to m from points of ℓ form an open segment of m.

10. ULTRA-IDEAL POINTS

Just as we defined an ideal point as a family of lines parallel in a given direction, so we shall define an *ultra-ideal point* to be a family of divergent lines (Fig. V-22).

Definition V-22

The family of all lines perpendicular to a given line is called an *ultra-ideal point*. The given line is called the *axis* or *carrier* of the ultra-ideal point.

In view of Theorem V-20, an ultra-ideal point Γ is obviously determined by any two lines of Γ, and through any point of the plane passes a unique member of Γ. An ultra-ideal point is determined also by its axis (which does not itself belong to Γ). The axis of an ultra-ideal point is the unique common perpendicular of any two of its lines. Thus, ultra-ideal points are in one-to-one correspondence with the lines of the plane, via $\Gamma \leftrightarrow$ axis of Γ.

By analogy with ideal points, we shall adopt the following suggestive terminology. If a line ℓ is a member of an ultra-ideal point Γ,

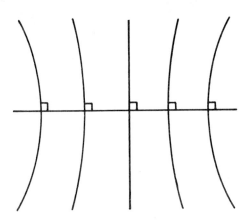

Figure V-22.

ULTRA-IDEAL POINTS

we shall say "ℓ contains Γ," "ℓ passes through Γ," or "Γ is on ℓ." Thus, two lines "meet" in an ultra-ideal point if and only if they both belong to the same family of divergent lines. Divergent lines "meet" in an ultra-ideal point. A (real) point A and an ultra-ideal point Γ determine a unique line, AΓ (See Exercise 1).

Exercises V-10

1. (a) Show that two distinct points, P and Q, determine a unique line in each of the following cases: (i) P real, Q ideal, (ii) P real, Q ultra-ideal, (iii) P and Q ideal. (b) If P and Q are both ultra-ideal, is there necessarily a line that "contains" them both? Explain. (c) What if P is ideal and Q is ultra-ideal?

2. It was mentioned earlier that congruence of segments and congruence of angles must be defined in a non-Euclidean way to make the Klein-Beltrami interpretation a model for Lobachevskian geometry. It turns out that with the proper definition of angle congruence, two "lines" ℓ and m are "perpendicular" if and only if the extension of one passes through the intersection of the tangents to the disk drawn at the extremities of the other (see the figure on this page). (A special case occurs when one of the "lines" is a diameter: in that case "lines" are Lobachevskian-perpendicular if they are so in the usual sense.)

   a. Given two divergent "lines" m and n in the Klein-Beltrami model, how would you draw their common perpendicular?

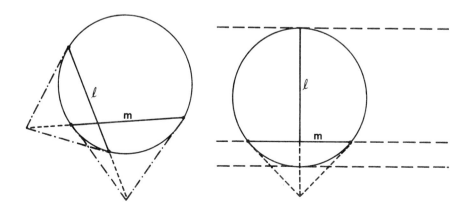

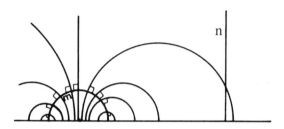

b. What do ultra-ideal points look like in the Klein-Beltrami model?

c. We saw in Exercise 1 of Section V-6 that ideal points are in one-to-one correspondence with points on the disk's boundary. Are ultra-ideal points in one-to-one correspondence with points external to the disk? (Are there exceptional cases?) How do you construct the family from the external point?

3. Interpret the various cases in Exercise 1 on the Klein-Beltrami Model. (Draw sketches.)

4. The figure above shows an ultra-ideal point in the Poincaré upper half-plane (recall congruence of "angles" means congruence of the angles formed by the tangents to the "sides"). "Line" m is the ultra-ideal point's axis. Sketch the ultra-ideal point whose axis is "line" n.

5. Sketch some ultra-ideal points in the Poincaré Disk Model.

11. SHEAVES OF LINES—THE FUNDAMENTAL CURVES

A real point also may be associated with a family of lines, namely the lines which pass through the point. The point is then called the *center* of the family. Since such families are in obvious 1-1 correspondence with real points, we may interchangeably think of a real point as either a point or as a family of lines with a common intersection. We therefore have three types of families of lines: (real) points, ideal points, and ultra-ideal points. In modern terminology, the word *sheaf* or *pencil* is often used in place of "family" (Fig. V-23).

In Euclidean geometry, the three perpendicular bisectors of the sides of a triangle meet in a real point. In Lobachevskian geometry, the three perpendicular bisectors belong to the same sheaf, which may be of any one of the three types.

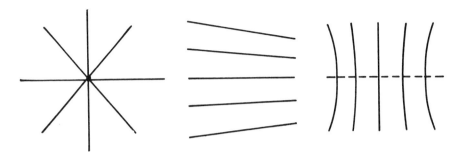

Figure V - 23.

Theorem V-23

The three perpendicular bisectors of the sides of a triangle are concurrent in either a real, ideal, or ultra-ideal point.

Proof. We have three cases.

Case 1: Suppose two of the perpendicular bisectors meet in a real point P. Then P is equidistant from all three vertices, and so P lies on the third bisector.

Case 2: Suppose, in $\triangle ABC$, two of the perpendicular bisectors, say $\overleftrightarrow{KL}$ of $\overline{AB}$ and $\overleftrightarrow{MN}$ of $\overline{BC}$, are divergent with common perpendicular $\overleftrightarrow{LN}$ (Fig. V-24). From the three vertices of $\triangle ABC$, drop perpendiculars $\overline{AA'}$, $\overline{BB'}$, and $\overline{CC'}$ to line $\overleftrightarrow{LN}$. The order of points on $\overleftrightarrow{LN}$ is as indicated in Figure V-24, since divergent lines cannot cross. Draw $\overline{AL}$ and $\overline{BL}$. $\triangle AKL \cong \triangle BKL$ and so AL = BL and $\measuredangle ALA' \cong \measuredangle BLB'$. Therefore $\triangle ALA' \cong \triangle BLB'$ and AA' = BB'. Similarly, BB' = CC'. Accordingly, figure ACC'A is a Saccheri quadrilateral, and so the perpendicular bisector of base $\overline{A'C'}$ is also the perpendicular bisector

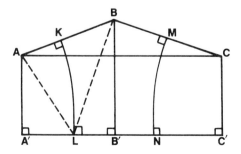

Figure V - 24.

of summit $\overline{AC}$. Thus all three perpendicular bisectors of $\triangle ABC$ are perpendicular to $\overleftrightarrow{LN}$.

Case 3: If two of the perpendicular bisectors are parallel, then all three must be parallel in pairs, for otherwise we would be in Case 1 or Case 2, a contradiction. We must prove they are all parallel in the same direction. First we shall show that all three perpendicular bisectors meet one of the sides of the triangle. Suppose, by relabeling if necessary, that $\angle ABC$ is the (or a) maximal angle of $\triangle ABC$. Within this angle, draw rays $\overrightarrow{BX}$ and $\overrightarrow{BY}$ so that $\angle ABX \cong \angle A$ and $\angle CBY \cong \angle C$ (Fig. V-25). By Theorem III-17, $\overrightarrow{BX}$ and $\overrightarrow{BY}$ meet $\overline{AC}$ in X and Y, respectively. (In the case of an isosceles or equilateral triangle, X and/or Y may coincide with C or A, respectively.) Since X and Y lie on the perpendicular bisectors of $\overline{AB}$ and $\overline{BC}$, all three perpendicular bisectors meet $\overline{AC}$.

Because all three perpendicular bisectors cross one of the sides of the triangle, two of them lie on opposite sides of the third (in effect, ruling out the formation of a "triangle" with three ideal vertices). The theorem now follows from the following.

Lemma V-24

If $\overleftrightarrow{AA'} \parallel \overleftrightarrow{CC'}$ and if $\overleftrightarrow{AA'}$ and $\overleftrightarrow{CC'}$ lie on opposite sides of $\overleftrightarrow{BB'}$, then all three lines are parallel in the same direction.

Proof. If P and Q are arbitrary points on $\overleftrightarrow{AA'}$ and $\overleftrightarrow{CC'}$, respectively, the $\overrightarrow{PQ}$ meets $\overleftrightarrow{BB'}$ in a point R. Any ray interior to $\angle A'PR$ meets $\overleftrightarrow{CC'}$ and so also $\overleftrightarrow{BB'}$. The angle criterion is established.

We now extend the notion of $P \triangleq Q$ introduced in Theorem V-6. Recall that there are three types of sheaves of lines: intersecting, parallel, and divergent.

Definition V-25

Let Σ be a sheaf of lines (Fig. V-26). Two distinct (real) points

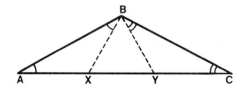

Figure V - 25.

SHEAVES OF LINES—THE FUNDAMENTAL CURVES

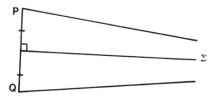

Figure V - 26.

P and Q are said to *correspond* relative to Σ if the perpendicular bisector of $\overline{PQ}$ belongs to Σ. We write P ≎ Q.

Equivalently, P ≎ Q (where P and Q are distinct from the center if Σ is an intersecting sheaf) if and only if segment $\overline{PQ}$ makes congruent angles with $\overrightarrow{PΣ}$ and $\overrightarrow{QΣ}$ (check this). Moreover, if Σ equals

a. A, a real point, then P ≎ Q if and only if P and Q are equidistant from A
b. Ω, an ideal point, then P ≎ Q if and only if the perpendicular bisector of $\overline{PQ}$ is the bisector of the strip between $\overrightarrow{PΩ}$ and $\overleftrightarrow{QΩ}$
c. Γ, an ultra-ideal point, then P ≎ Q if and only if P and Q are equidistant from and on the same side of the axis of Γ

If we agree to say that any point corresponds to itself, then correspondence relative to a fixed sheaf is an equivalence relation. That is, if P, Q, and R are any three (real) points, then, relative to Σ,

a. P ≎ P (reflexivity)
b. P ≎ Q implies Q ≎ P (symmetry)
c. P ≎ Q and Q ≎ R implies P ≎ R (transitivity)

Symmetry and reflexivity are obvious. Transitivity follows from Theorem V-23, unless P, Q, and R are collinear, in which case Σ is ultra-ideal and P, Q, and R all lie on the axis of Σ. In this case P ≎ R is clear.

Definition V-26
 Let Σ be a sheaf of lines and let P be a point (other than the center in the case where Σ is a sheaf of intersecting lines). Then the set of all points that correspond to P relative to Σ is called

a. a *circle*, if Σ is a real point
b. a *limiting curve* or *horocycle*, if Σ is an ideal point
c. an *equidistant curve*, if Σ is an ultra-ideal point

By the transitivity of correspondence, any two points on one of these curves correspond relative to the sheaf used to define the curve. Evidently, a circle is the set of all points at a fixed distance from the center of a sheaf of intersecting lines. An equidistant curve consists of all the points which are on one side of and at a fixed distance from the axis of an ultra-ideal point. Of these three types of curve, only circles are closed curves.

Theorem V-27

No three distinct points of a circle, limiting curve, or equidistant curve with positive distance are collinear.

Proof. If three points A, B, and C on one of these curves are collinear, then the perpendicular bisectors of $\overline{AB}$ and $\overline{BC}$ are divergent lines. Σ is in this case an ultra-ideal point, and A, B, and C lie on its axis. The curve is therefore an equidistant curve of distance zero.

Exercises V-11

1. How many different types of sheaf are there in Euclidean geometry?

2. (a) Show that in Euclidean geometry, only Case 1 of Theorem V-23 can occur. (b) Conversely, show that if only Case 1 can occur, then Euclid's postulate holds. (Hint: see Section IV-7.)

3. Some authors define an equidistant curve to be the set of all points at a fixed distance from a given line. This consists of two curves or *branches*, one on each side of the line. Prove that the line joining a point on one branch of an equidistant curve to a point on the other is bisected by the curve's axis.

12. LIMITING CURVES

Consider one of the limiting curves S determined by an ideal point Ω (Fig. V-27). By analogy with circles, we shall refer to Ω as the *center* of S, and if A is any point of S, we shall call $\overleftrightarrow{A\Omega}$ a *radius* or *axis* of S. Two different limiting curves with the same center are called *concentric*.

If A and B are on S, then $\angle AB\Omega \cong \angle BA\Omega = \Pi(AB/2)$. Conversely, if A is any point of S and if ray $\overrightarrow{AB}$ is such that $\angle BA\Omega$ is acute,

LIMITING CURVES

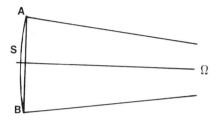

Figure V - 27.

then ray $\overrightarrow{AB}$ will meet S in a unique point B satisfying $\Pi(AB/2) = \measuredangle BA\Omega$. Accordingly, of the non-radial lines through A, all but one meet S in a second point. The exception is the line (called the *tangent* at A) which is perpendicular to the radius $\overleftrightarrow{A\Omega}$.

We next show that all limiting curves are congruent in the sense of the following definition.

Definition V-28

Two curves are *congruent* if there is a one-to-one correspondence between their points under which each chord of one curve corresponds to a congruent chord of the other.

Theorem V-29

Any two limiting curves are congruent.

Proof. Let S and S' be limiting curves with centers Ω and Ω', respectively (Fig. V-28). Choose points A on S and A' on S'. Randomly specify a one-to-one correspondence between the two half-planes

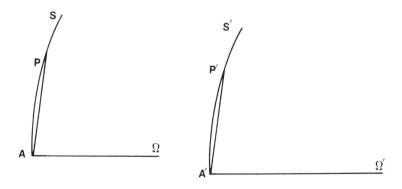

Figure V - 28.

determined by line $\overleftrightarrow{A\Omega}$ and the two half-planes determined by $\overleftrightarrow{A'\Omega'}$. Then to each point P of S (other than A) we make correspond the unique point P' of S' which lies in the corresponding half-plane and satisfies $\measuredangle PA\Omega \cong \measuredangle P'A'\Omega'$. Then, since $\Pi(PA/2) = \measuredangle PA\Omega \cong \measuredangle P'A'\Omega' = \Pi(P'A'/2)$ and since Π is one-to-one, we have PA = P'A'.

Similarly, if Q on S and Q' on S' are a second pair of corresponding points, we have QA = Q'A' and $\measuredangle PAQ \cong P'A'Q'$. Therefore $\triangle PAQ \cong \triangle P'A'Q'$ and so the corresponding chords $\overline{PQ}$ and $\overline{P'Q'}$ are congruent. Therefore S and S' are congruent.

It is clear from the preceding that on any two limiting curves (distinct or not), congruent limiting arcs subtend congruent chords and vice versa.

If A, B, and C lie on a limiting curve S with center Ω, we shall say that B is *between* A and C if and only if $\overleftrightarrow{A\Omega}$ and $\overleftrightarrow{C\Omega}$ are on opposite sides of $\overleftrightarrow{B\Omega}$ (see the last paragraph of Section V-4).

Having defined congruence and betweenness for limiting arcs, it is then a simple matter to define inequality of limiting arcs: arc $\widehat{AB}$ < arc $\widehat{CD}$ if and only if chord $\overline{AB}$ < chord $\overline{CD}$. The reader may check that Hilbert's order postulates II, 1, 2, 3 and congruence postulates III, 1, 2, 3 hold for points and arcs of limiting curves. It is now possible, by choosing a convenient limiting arc as a unit, to assign a measure of length to limiting arcs.

In addition, Dedekind's postulate holds for limiting arcs. For, if $\widehat{AB}$ is a limiting arc, then the radius through any of its points P will intersect chord $\overline{AB}$ in a unique point P'. This gives an order-preserving correspondence of the points of arc $\widehat{AB}$ with the points of chord $\overline{AB}$. Consequently, Dedekind's postulate for segments implies its analog for limiting arcs. These results will be needed in Chapter VI.

Exercises V-12

1. How many limiting curves pass through two distinct real points? Why?

2. Show that there are infinitely many equidistant curves passing through two distinct points.

3. Let m and n be divergent lines with common perpendicular $\overleftrightarrow{MN}$, where M is on m and N is on n. Let A ($\neq$ M) be a point of m. Show that if B is the point on m such that AM = MB and M is between A and B, then A and B are equidistant from n.

LIMITING CURVES

4. Let S be an equidistant curve (Definition V-26). Let ℓ be the axis of the associated ultra-ideal point Γ. For any point A of S, $\overleftrightarrow{A\Gamma}$ denotes the perpendicular to ℓ through A. The ray $\overrightarrow{A\Gamma}$ is directed from A toward ℓ. Prove the following.

   a. If A and B are any two distinct points of S, then angles $\angle BA\Gamma$ and $\angle AB\Gamma$ are congruent and acute
   b. If A is any point of S and if ray $\overrightarrow{AB}$ is such that $\angle BA\Gamma$ is acute, then ray $\overrightarrow{AB}$ need not intersect S in a point other than A
   c. The line t through A which is perpendicular to $\overleftrightarrow{A\Gamma}$ does not meet S except in A. (t is called the *tangent* to S at A.)
   d. The only rays through A which make acute angles with $\overrightarrow{A\Gamma}$, do not meet S in another point, and do not meet ℓ, are the parallels to ℓ through A
   e. Let A be a fixed point of S and let B on S be a variable point. Give an informal argument supporting the fact that as B moves along S toward A, $\angle BA\Gamma$ approaches a right angle. (This justifies the name "*tangent*" given to the perpendicular to $\overleftrightarrow{A\Gamma}$ through A.)

5. Prove that the line joining a point on one branch of an equidistant curve to a point on the other makes congruent alternate interior angles with the (tangents to the) curve (cf. Exercise 3 of Section V-11).

6. (a) Prove that any two equidistant curves with the same distance are congruent. (b) Where does your proof break down if the two equidistant curves do not have the same distance?

7. Let C be a variable point on ray $\overrightarrow{P\Omega}$. Consider the circle centered at C and passing through P. Let A be a point of this circle which does not lie on $\overleftrightarrow{PC}$. Consider what happens as C moves away from P in the direction of Ω. Each position of C gives a different circle through P, and as C moves farther from P these circles become larger and larger. For each position of C, choose A so that the length of chord PA remains constant. (The figure on p. 198 shows four positions of C along with the corresponding positions of A.) Explain why, as C recedes indefinitely toward Ω, angles $\angle PAC$ and $\angle CPA$ approach $\Pi(AP/2)$ and A approaches a point on the horocycle through P with center Ω; cf. Exercise 7 of Section V-7. (Note: the preceding

indicates that a horocycle may be viewed as the limit of circles, hence the name "limiting curve." Although circles are closed curves, you should not conclude that horocycles are closed. In the above construction, as $C \to \Omega$, each point A of the circle that is not on diameter $\overleftrightarrow{PC}$ moves along a circle (of radius PA) toward a limit. But the point diametrically opposite P on the variable circle does not approach any (real) point as $C \to \Omega$. The horocycle does not contain a point on ray $\overrightarrow{P\Omega}$ other than P itself. It seems strange to us that a curve can continually bend (with constant curvature) and yet not be closed, but this is because we are forming mental pictures on a flat (Euclidean) plane. In Euclidean geometry, the limit of the circle centered at C would be the straight line perpendicular to $\overleftrightarrow{P\Omega}$ at P.)

8. As in Exercise 7, let C be a variable point on $\overrightarrow{P\Omega}$. Let ℓ be the line through C perpendicular to $\overleftrightarrow{P\Omega}$. Let S be the equidistant curve passing through P with axis ℓ. When C coincides with P, S is a line (equidistant curve of distance zero). However, as C and ℓ move away from P, S becomes "more curved," and approaches the horocycle through P with radius $\overrightarrow{P\Omega}$. Give a plausability argument for this similar to that used for Exercise 7.

9. We have seen that each radius of a horocycle intersects the horocycle orthogonally. As we sometimes say, a horocycle is an *orthogonal trajectory* of a family of parallel lines. Similarly, a circle is an orthogonal trajectory of its radii, and an equidistant curve is an orthogonal trajectory of a family of divergent lines. This orthogonality enables us to sketch these curves in models for Lobachevskian

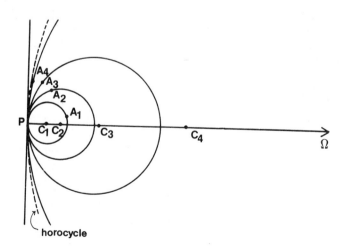
horocycle

LIMITING CURVES

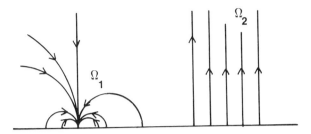

geometry in which angles are measured in the usual Euclidean way.

a. The above figure shows some of the "lines" of two ideal points, Ω_1 and Ω_2 in the Poincaré Upper Half-Plane Model. The arrows indicate directions of parallelism. Sketch some of the horocycles associated with these ideal points. Describe all horocycles in this model.
b. In the Upper Half-Plane Model, sketch a family of intersecting "lines" and some of the "circles" centered at their common "point." (It turns out that "circles" in this model are Euclidean circles, but their Lobachevskian "centers" are not their Euclidean centers.)
c. Sketch several equidistant curves for the ultra-ideal point whose axis is m in the figure of Exercise 4, Section V-10. (These turn out to be Euclidean circles too.) Sketch some equidistant curves for the ultra-ideal point with axis n in the figure on p. 190.
d. Sketch some circles, horocycles, and equidistant curves in the Poincaré disk model. (They are all Euclidean circles.)

13. CONCENTRIC HOROCYCLES

Let $\widehat{AB}$ and $\widehat{A'B'}$ be concentric limiting arcs, where $\overleftrightarrow{AA'}$ and $\overleftrightarrow{BB'}$ are radii, as in Figure V-29. The following lemma asserts that AA' = BB'.

Lemma V-30

Segments of radii intercepted between two concentric horocycles are congruent.

Proof. Refer to Figure V-29. The bisector of the strip between $\overleftrightarrow{AA'}$ and $\overleftrightarrow{BB'}$ will intersect chords $\overline{AB}$ and $\overline{A'B'}$ perpendicularly at their midpoints, C and C', respectively. Draw $\overline{AC'}$ and $\overline{BC'}$. Then $\triangle ACC' \cong \triangle BCC'$ and $\triangle AA'C' \cong \triangle BB'C'$, hence AA' = BB'.

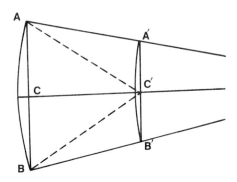

Figure V - 29.

We shall refer to AA' as the *distance* between the concentric limiting arcs. Note that the bisector of the strip between $\overleftrightarrow{AA'}$ and $\overleftrightarrow{BB'}$ also bisects arcs $\widehat{AB}$ and $\widehat{A'B'}$. From this observation, the reader can easily verify the following.

Lemma V-31

Let $\widehat{AB}$ and $\widehat{A'B'}$ be concentric limiting arcs with center Ω. Suppose $\widehat{AB}$ is divided into n congruent parts by the points P_0 (= A), $P_1, P_2, \ldots, P_n$ (= B), and that the radii drawn through these points intersect $\widehat{A'B'}$ in Q_0 (= A'), $Q_1, Q_2, \ldots, Q_n$ (= B'), respectively. Then the latter divide $\widehat{A'B'}$ into n congruent parts.

This leads to the following important result.

Theorem V-32

Let A, B, and C be three distinct points on a limiting curve S with center Ω. Let A', B' and C' be the intersections of $\overleftrightarrow{A\Omega}$, $\overleftrightarrow{B\Omega}$, and $\overleftrightarrow{C\Omega}$ with a concentric horocycle S'. Then $\widehat{AB}/\widehat{A'B'} = \widehat{AC}/\widehat{A'C'}$.

Proof. First, suppose that $\widehat{AB}$ and $\widehat{AC}$ are *commensurable*. This means that their ratio is a rational number so that there is a unit limiting arc u and positive integers m and n such that u is contained in $\widehat{AB}$ m times and in $\widehat{AC}$ n times. Thus, we may mark points on these arcs so that $\widehat{AB}$ is divided into m congruent parts and $\widehat{AC}$ is divided into n congruent parts, and adjacent points are separated by arcs congruent to u. By drawing the radii through these points and applying the Lemma, we obtain $\widehat{AB}/\widehat{AC} = m/n = \widehat{A'B'}/\widehat{A'C'}$, or $\widehat{AB}/\widehat{A'B'} = \widehat{AC}/\widehat{A'C'}$.

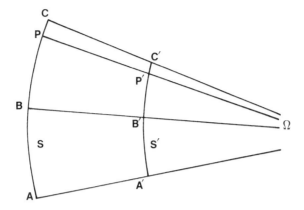

Figure V - 30.

In the case where $\widehat{AB}$ and $\widehat{AC}$ are *incommensurable*, we must use a limit argument. Choose as a unit arc, any limiting arc u which is contained in $\widehat{AB}$ an integral number of times, m. Then $\widehat{AC}$ will consist of a limiting arc $\widehat{AP}$, congruent to an integral multiple of u, plus a remainder $\widehat{PC}$, which is smaller than u (see Fig. V-30). Let $\overleftrightarrow{P\Omega}$ intersect S' in P'. By the commensurable case, $\widehat{AB}/\widehat{A'B'} = \widehat{AP}/\widehat{A'P'}$. If we now let $m \to \infty$, then u and hence the remainders $\widehat{PC}$ and $\widehat{P'C'}$ will approach zero. Thus, $\widehat{AP}$ and $\widehat{A'P'}$ will approach $\widehat{AC}$ and $\widehat{A'C'}$, respectively, and the result follows.

Consider now two pairs of concentric limiting arcs: $\widehat{AB}$ and $\widehat{A'B'}$ with center Ω_1; and $\widehat{CD}$ and $\widehat{C'D'}$ with center Ω_2 (Fig. V-31). Suppose $AA' = CC'$. We claim that $\widehat{AB}/\widehat{A'B'} = \widehat{CD}/\widehat{C'D'}$.

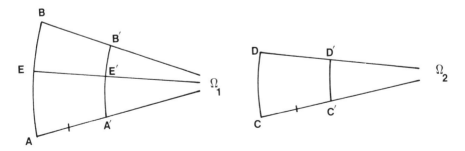

Figure V - 31.

On $\stackrel{\frown}{AB}$, or on $\stackrel{\frown}{AB}$ extended through B, mark the point E such that $\stackrel{\frown}{AE} = \stackrel{\frown}{CD}$ and let E' be the intersection of the radius $\overleftrightarrow{E\Omega_1}$ with $\stackrel{\frown}{A'B'}$ (or its extension). Then $\stackrel{\frown}{A'E'} = \stackrel{\frown}{C'D'}$ (see Exercise 1), and so figures AEE'A' and CDD'C' are congruent. Consequently, $\stackrel{\frown}{AB}/\stackrel{\frown}{A'B'} = \stackrel{\frown}{AE}/\stackrel{\frown}{A'E'} = \stackrel{\frown}{CD}/\stackrel{\frown}{C'D'}$.

Theorem V-32 and the preceding argument may be summarized as follows.

Theorem V-33

The ratio of two concentric limiting arcs intercepted between radii depends only on the distance between the arcs, and not on their size or location in the plane.

The trigonometric relations of Lobachevskian geometry ultimately stem from the following remarkable theorem.

Theorem V-34

The ratio of concentric limiting arcs s_1 and s_2 ($s_1 > s_2$) intercepted between two radii is given by

$$s_1/s_2 = e^{x/k}$$

where x is the distance between the arcs and k is a positive constant.

Proof. Consider any three concentric limiting arcs, s_1, s_2, and s_3, intercepted between two radii. Assume $s_1 > s_2 > s_3$. In view of Theorem V-12, the length of concentric arcs decreases in the direction of parallelism. The situation is depicted in Figure V-32, where x is the distance between s_1 and s_2, while y is the distance between s_2 and s_3.

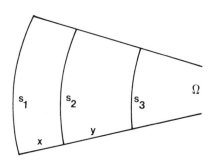

Figure V - 32.

CONCENTRIC HOROCYCLES

In view of the preceding theorem, we may write

$$s_1/s_2 = f(x), \quad s_2/s_3 = f(y), \quad s_1/s_3 = f(x+y)$$

where f is a real-valued function that is clearly positive and increasing. These equations yield the functional relation

$$f(x+y) = f(x)\,f(y) \quad \text{for } x,y \geq 0$$

f is known to be continuous (assume this). We can readily show that a function with all these properties is an exponential function of the form a^x, where $a > 1$. This may be done as follows.

Setting $y = 0$ in the above relation, we deduce $f(0) = 1$. By induction on the number of summands, $f(nx) = f(x)^n$, for any $x \geq 0$ and any positive integer n. In particular, since $f(1) = f[n(1/n)] = f(1/n)^n$, we deduce that for any non-negative rational number $r = m/n$,

$$f(r) = f(m/n) = f(1/n)^m = f(1)^{m/n}$$

If we denote $f(1)$ by a, this becomes

$$f(r) = a^r, \quad r \text{ rational} \geq 0$$

By continuity, $f(x) = a^x$ for all real $x \geq 0$. Letting $k = 1/(\ln a)$, we obtain (since $a^x = (e^{\ln a})^x = e^{x/k}$)

$$s_1/s_2 = e^{x/k}$$

as required. $k > 0$ since $a > 1$ (f is increasing).

Since x is a length in the above formula, so is k. It is in fact the distance between concentric limiting arcs (intercepted between radii) whose ratio equal $e = 2.71828\ldots$. This distance k is called the *radius of curvature* or *space constant* and is an example of an absolutely determined length. We shall see later that it is the analogue of the radius of the sphere in spherical geometry. Note that although as a *length* k is absolutely determined, its *numerical* value will depend on an arbitrary choice of a unit segment. Although we shall not do so here, some authors choose k itself as the unit segment, so that the numerical value assigned to k is one.

The unexpected appearance of the exponential function at this point is a tremendous breakthrough, for now the way has been paved for analysis to enter upon the scene. As you will discover, we have found the key to the trigonometry of the Lobachevskian plane, which will be developed next, in Chapter VI.

Exercises V-13

1. In Figure V-31, show that $\widehat{A'E'} \cong \widehat{C'D'}$. (Hint: draw $\overline{AE'}$ and $\overline{CD'}$. Use congruent triangles to show $\overline{A'E'} \cong \overline{C'D'}$.)

2. Which of the theorems and lemmas of this section hold true for circles in Euclidean geometry and which do not?

VI

THE TRIGONOMETRIC FORMULAS

> *Nothing puzzles me more than time and space; and yet nothing troubles me less, as I never think about them.*
> —Charles Lamb

> *Geometry, which is the only science it hath pleased God hitherto to bestow on mankind.*
> —Thomas Hobbes

1. PERPENDICULAR LINES AND PLANES

In deriving the trigonometric formulas of the non-Euclidean plane, we shall follow a method similar to that used independently by Lobachevsky and Bolyai. Since this method employs the properties of a certain curved surface in space—the horosphere—we shall need a few facts from 3-dimensional geometry. None of the proofs presented in Section 1 make use of Euclid's parallel postulate or its equivalents. Hence all the results in this section are absolute, i.e., they are equally true in Euclidean as well as Lobachevskian solid geometry. Since many of these results may be familiar to you, or at least readily believable on the basis of your intuition, you may wish to omit their proofs on a first reading, and only briefly scan the statements.

Throughout, ℓ, m, n, etc., denote lines, while α, β, γ, etc., denote planes. As before, upper case English letters usually will be used to denote points. The standard symbol ∩ indicates intersection. Thus, ℓ ∩ m and ℓ ∩ α generally indicate points; α ∩ β, a line. The plane determined by a line ℓ and a point P not on ℓ will be denoted Pℓ.

Definition VI-1
    Line ℓ is said to be *perpendicular* to plane α if ℓ intersects α in a point, P, and if ℓ is perpendicular to every line of α that passes through P.

Lemma VI-2
    If a line is perpendicular to two lines of a given plane at their point of intersection, then it is perpendicular to the given plane.
    Proof. In Figure VI-1, lines m and n of plane α intersect at P, $\overleftrightarrow{AP} \perp m$, and $\overleftrightarrow{AP} \perp n$. According to Definition VI-1, we must show that $\overleftrightarrow{AP}$ is perpendicular to any third line of α, say $\overleftrightarrow{PD}$, which also passes through P.
    To this end, choose points B on m and C on n on opposite sides of $\overleftrightarrow{PD}$. Then $\overline{BC}$ meets $\overleftrightarrow{PD}$ in a point, which we may as well assume is the one labeled D. Extend $\overline{AP}$ through P to A' so that PA = PA', and draw all remaining lines of the figure. The remaining steps of the proof are as follows.

    a. $\overleftrightarrow{BP}$ and $\overleftrightarrow{CP}$ are perpendicular bisectors of $\overline{AA'}$
    b. △ABC ≅ △A'BC (SSS)
    c. ∡ABD ≅ ∡A'BD

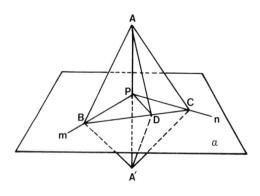

Figure VI - 1.

PERPENDICULAR LINES AND PLANES

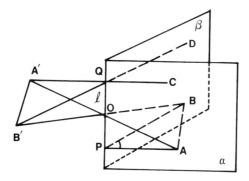

Figure VI - 2.

d. $\triangle ABD \cong \triangle A'BD$ (SAS)
e. $AD = A'D$
f. $\overleftrightarrow{DP}$ is a perpendicular bisector of $\overline{AA'}$

Accordingly, $\overleftrightarrow{AP} \perp \overleftrightarrow{PD}$, as was to be shown.

Definition VI-3
  The figure consisting of a line and two non-coplanar half-planes having the line as common edge is called a *dihedral angle*.
  In Figure VI-2, a dihedral angle is formed by the half-planes α and β. P is any point of $\ell = \alpha \cap \beta$. Through P, in α and in β, respectively, the rays $\overrightarrow{PA}$ and $\overrightarrow{PB}$ are drawn perpendicular to ℓ. $\angle APB$ is called a *plane angle* of the dihedral angle.

Lemma VI-4
  All plane angles of a given dihedral angle are congruent.
  Proof. In Figure VI-2, $\angle BPA$ and $\angle DQC$ are plane angles of the dihedral angle formed by half-planes α and β. This means that $\overleftrightarrow{AP}$, $\overleftrightarrow{BP}$, $\overleftrightarrow{CQ}$, and $\overleftrightarrow{DQ}$ are all perpendicular to $\overleftrightarrow{PQ}$.
  Let O be the midpoint of segment $\overline{PQ}$. In α, draw $\overline{AO}$ and extend it through O to A' so that $OA = OA'$. In β, draw $\overline{BO}$ and extend it through O to B' so that $OB = OB'$. Draw the remaining lines in Figure VI-2. Then

a. $\triangle AOP \cong \triangle A'OQ$ and $\triangle BOP \cong \triangle B'OQ$ (SAS)
b. $\overleftrightarrow{A'Q} \perp \overleftrightarrow{PQ}$, $\overleftrightarrow{B'Q} \perp \overleftrightarrow{PQ}$

c. A', Q, C are collinear; B', Q, D are collinear
d. AP = A'Q, BP = B'Q
e. △A'OB' ≅ △AOB (SAS)
f. AB = A'B'
g. △A'B'Q ≅ △ABP (SSS)
h. ∡APB ≅ ∡A'QB ≅ ∡DQC, as claimed

By virtue of Lemma VI-4, we may unambiguously assign, as the measure of a dihedral angle, the measure of any of its plane angles.

Definition VI-5
   Two half-planes with common edge will be called *perpendicular* if the dihedral angle they form has plane angles which are right. Two intersecting planes will be called *perpendicular* if any one (and hence all four) of the dihedral angles they form has (have) plane angles which are right.

Theorem VI-6
   If a line is perpendicular to a given plane, then any plane through that line is perpendicular to the given plane.
   Proof. In Figure VI-3, $\ell \perp \alpha$ and β passes through ℓ. In α, draw $m \perp (\alpha \cap \beta)$. Since $\ell \perp m$ (Definition VI-1), α and β form a dihedral angle with a right plane angle. In other words, $\alpha \perp \beta$.

Lemma VI-7
   If a line lying in one of two perpendicular planes is perpendicular to their line of intersection, then it is perpendicular to the other plane.

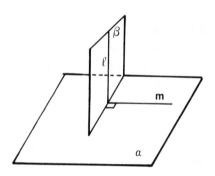

Figure VI - 3.

PERPENDICULAR LINES AND PLANES

Proof. Consider Figure VI-3 again, where we assume this time that $\alpha \perp \beta$ and $\ell \perp (\alpha \cap \beta)$. m is drawn as before. Since ℓ and m together form a plane angle of a right dihedral angle, $\ell \perp$ m. Now apply Lemma VI-2 to deduce $\ell \perp \alpha$.

Theorem VI-8
Through a given point, a unique line may be drawn perpendicular to a given plane.

Proof. Let P be the given point and α the given plane. The proof is essentially the same whether P lies on α (Fig. VI-4a) or not (Fig. VI-4b).

Let C be the foot of the perpendicular dropped from P to an arbitrary line $\overleftrightarrow{AB}$ in α. In any plane through $\overleftrightarrow{AB}$ other than plane PAB, draw $\overleftrightarrow{CD} \perp \overleftrightarrow{AB}$. [In case (b), $\overleftrightarrow{CD}$ may be drawn in α.] Then $\overleftrightarrow{PC}$ and $\overleftrightarrow{CD}$ determine a plane β perpendicular to α (Lemma VI-2, Theorem VI-6). Let $\overleftrightarrow{CE} = \alpha \cap \beta$. In plane β, draw the perpendicular to $\overleftrightarrow{CE}$ through P. This line will be perpendicular to α (Lemma VI-7).

If there were two distinct perpendiculars to α through P, they would determine a plane γ in which there would be two lines through P perpendicular to $\alpha \cap \gamma$, an impossibility.

Projections

Let ℓ be any line not perpendicular to plane α. From any point P on ℓ, drop the perpendicular $\overline{PP'}$ to α, and let β = plane P'ℓ. Then $\beta \perp \alpha$ (Fig. VI-5).

If Q is any other point of ℓ, let Q' be the foot of the perpendicular dropped in β from Q to $\alpha \cap \beta$. By Lemma VI-7, $\overline{QQ'} \perp \alpha$. Since Q was arbitrarily chosen, it follows that the feet of all

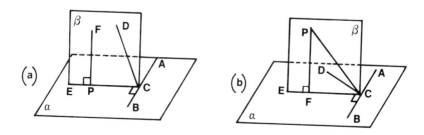

Figure VI-4.

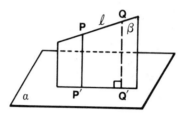

Figure VI - 5.

perpendiculars dropped to α from points of ℓ are collinear, forming a (proper) subset of line $\alpha \cap \beta$ (see Exercise 1). This subset is called the *projection* of ℓ onto α, denoted $\text{proj}_\alpha \ell$.

Definition VI-9

Let ℓ be a line not perpendicular to plane α; $\text{proj}_\alpha \ell$ is the set consisting of the feet of all the perpendiculars dropped from points of ℓ to plane α.

We shall occasionally allow ourselves a little sloppiness in terminology and say that a certain line m is the projection of another line ℓ onto a plane α, even though $\text{proj}_\alpha \ell$ is only a subset of m.

If ℓ intersects α, then either of the two acute vertical angles formed by ℓ and $\text{proj}_\alpha \ell$ may be taken as a measure of the inclination of ℓ to α.

Lemma VI-10

Suppose line ℓ meets plane α, but not perpendicularly, in a point A. Then the acute angle between ℓ and $\text{proj}_\alpha \ell$ is less than the angle between ℓ and any other line of α that passes through A.

Proof. Let Q be the foot of the perpendicular dropped to α from any point P on ℓ (P ≠ A). Then line $\overleftrightarrow{AQ}$ contains $\text{proj}_\alpha \ell$. Let $\overleftrightarrow{AS}$ be a line of α distinct from $\overleftrightarrow{AQ}$ (Fig. VI-6). We must prove ∡PAR > ∡PAQ.

If ∡PAR is right or obtuse, this is obvious, since ∡PAQ is acute. Hence, we can proceed to the case in which ∡PAR is acute.

Drop $\overline{PR} \perp \overleftrightarrow{AS}$ and draw $\overline{QR}$. Then ∡PQR is right, and so PR > PQ. Comparing the right triangles △PAR and △PAQ, we see that they share a common hypotenuse, but that leg PR > leg PQ. It

PERPENDICULAR LINES AND PLANES 211

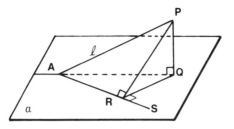

Figure VI - 6.

follows from absolute geometry that ∢PAR > ∢PAQ (see Exercise 3).

Lemma VI-11

If a plane is perpendicular to each of two intersecting planes, then it is perpendicular to their line of intersection.

Proof. Let plane γ be perpendicular to planes α and β, and let $\ell = \alpha \cap \beta$. Suppose, contrary to what we wish to prove, either that ℓ and γ do not intersect (Fig. VI-7a), or that ℓ and γ do intersect, but not perpendicularly (Fig. VI-7b).

In the former case, let P be any point on $\overleftrightarrow{\ell}$. In the latter, let $P = \ell \cap \gamma$. In either case, in plane α, draw $\overleftrightarrow{PQ}$ perpendicular to $(\alpha \cap \gamma)$. By Lemma VI-7, $\overleftrightarrow{PQ} \perp \gamma$. Likewise, in β, draw $\overleftrightarrow{PR}$ perpendicular to $(\beta \cap \gamma)$. Then $\overleftrightarrow{PR} \perp \gamma$. Since, according to our supposition, $\overleftrightarrow{PR}$ and $\overleftrightarrow{PQ}$ must be distinct lines, we have a contradiction to the uniqueness asserted in Theorem VI-8. Consequently, our supposition is untenable, and $\ell \perp \gamma$.

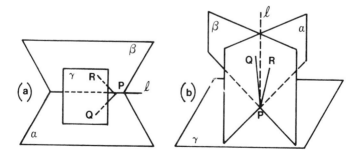

Figure VI - 7.

Lemma VI-12

Let line ℓ intersect plane α nonorthogonally in point P, and suppose m is a line of α which passes through P. Then $\ell \perp m$ if and only if $\text{proj}_\alpha \ell \perp m$.

Proof. Let β be the plane of ℓ and $\text{proj}_\alpha \ell$. $\beta \perp \alpha$. In β erect the line ℓ' perpendicular to $\text{proj}_\alpha \ell$. Then $\ell' \perp \alpha$, and so $\ell' \perp m$ (Fig. VI-8).

Now if either one of ℓ or $\text{proj}_\alpha \ell$ is perpendicular to m, then (since $\ell' \perp m$ as well) it follows that $\beta \perp m$, by Lemma VI-2. Consequently, the other of ℓ or $\text{proj}_\alpha \ell$ must be perpendicular to m also (Definition VI-1).

Exercises VI-1

1. Let ℓ be a line not perpendicular to plane α. Show (a) if ℓ and $\text{proj}_\alpha \ell$ intersect, then $\text{proj}_\alpha \ell$ is an open segment (cf. Theorem V-9); (b) if ℓ and $\text{proj}_\alpha \ell$ are parallel, then $\text{proj}_\alpha \ell$ is an open ray (cf. Exercise 6 of Section V-5); (c) if ℓ and $\text{proj}_\alpha \ell$ are divergent, then $\text{proj}_\alpha \ell$ is an open segment (cf. Exercise 4 of Section V-9).

2. (a) Consider the angle formed by two great circle arcs $\widehat{AB}$ and $\widehat{AC}$ on a sphere. Each of these arcs lies in a unique plane through the sphere's center. What is the relationship between the dihedral angle formed by these two planes and the spherical angle ∡BAC? Explain. (b) Prove that if three planes have a unique point in common, then the sum of the three dihedral angles formed by these planes in pairs exceeds two right angles.

3. Let $\triangle ABC$ and $\triangle A'B'C'$ have right angles at C and C', AB = A'B', and BC < B'C'. Prove ∡BAC < ∡B'A'C'. (Since we are in the context of absolute geometry here, you may not use the Pythagorean Theorem.)

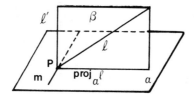

Figure VI - 8.

2. PARALLEL LINES AND PLANES

We henceforth assume Lobachevsky's postulate (Section IV-2).

Definition VI-13

Two distinct lines in space, $\overleftrightarrow{AA'}$ and $\overleftrightarrow{BB'}$, are said to be *parallel* (in the direction $\overrightarrow{AA'}$) if they lie in a single plane and, in that plane, (a) and (b) of Definition V-5 hold.

As in Section V-3, the direction of parallelism will be indicated by the ordering of the letters in $\overleftrightarrow{AA'}$ and $\overleftrightarrow{BB'}$. Parallelism of lines in space is symmetric: $\overleftrightarrow{AA'} \parallel \overleftrightarrow{BB'}$ implies $\overleftrightarrow{BB'} \parallel \overleftrightarrow{AA'}$. (This follows immediately by applying Theorem V-7 in the plane containing the two given lines.) As in two dimensions, given a point P not on a line ℓ, there exists one line through P parallel to ℓ in each direction. Transitivity of parallelism will be deduced from the following lemma.

Lemma VI-14

If two intersecting planes pass through two given parallel lines, then their line of intersection is parallel to both of the lines in their direction of parallelism.

Proof. In Figure VI-9, parallel lines $\overleftrightarrow{AA'}$ and $\overleftrightarrow{BB'}$ lie respectively in the distinct planes α and β, which intersect in line $\overleftrightarrow{CC'}$. We shall prove that $\overleftrightarrow{CC'} \parallel \overleftrightarrow{AA'}$.

First of all, $\overleftrightarrow{CC'}$ and $\overleftrightarrow{AA'}$ cannot intersect, for if they had a point D in common, then we would have

β = plane BB'CC' = plane BB'D = plane BB'AA',

which would imply that $\overleftrightarrow{AA'}$, $\overleftrightarrow{BB'}$, and $\overleftrightarrow{CC'}$ all lie in one plane, impossible since α and β were assumed distinct.

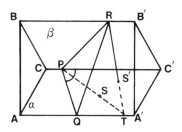

Figure VI - 9.

It remains to establish the angle criterion. Choose points P on $\overleftrightarrow{CC'}$, Q on $\overleftrightarrow{AA'}$, and R on $\overleftrightarrow{BB'}$. We must show that any ray $\overrightarrow{PS}$ lying within angle ∢C'PQ (opening in the direction of parallelism $\overrightarrow{AA'}$) meets $\overrightarrow{QA'}$. Consider plane RPS, which intersects plane AA'BB' in a line $\overrightarrow{RS'}$, which passes within ∢QRB'. Since $\overleftrightarrow{BB'} \parallel \overleftrightarrow{AA'}$, $\overrightarrow{RS'}$ must meet $\overrightarrow{AA'}$ in a point T, through which $\overrightarrow{PS}$ must pass (because T, like $\overrightarrow{PS}$, lies on plane RPS as well as on α).

Hence the angle criterion holds, and $\overleftrightarrow{CC'} \parallel \overleftrightarrow{AA'}$. Similarly, $\overleftrightarrow{CC'} \parallel \overleftrightarrow{BB'}$.

Theorem VI-15

If $\overleftrightarrow{AA'} \parallel \overleftrightarrow{BB'}$ and $\overleftrightarrow{BB'} \parallel \overleftrightarrow{CC'}$, then $\overleftrightarrow{AA'} \parallel \overleftrightarrow{CC'}$ (assuming $\overleftrightarrow{AA'}$ and $\overleftrightarrow{CC'}$ are distinct).

Proof. Suppose $\overleftrightarrow{AA'} \parallel \overleftrightarrow{BB'}$ and $\overleftrightarrow{BB'} \parallel \overleftrightarrow{CC'}$. We shall prove $\overleftrightarrow{AA'} \parallel \overleftrightarrow{CC'}$. If all three lines are coplanar, this follows from Theorem V-8. Otherwise, let P be any point of $\overleftrightarrow{CC'}$ and consider planes PAA' and PBB'. By the preceding lemma, these planes intersect in a line ℓ which passes through P and is parallel to ($\overrightarrow{AA'}$ and) $\overrightarrow{BB'}$. However, $\overleftrightarrow{CC'}$ also passes through P and is parallel to $\overrightarrow{BB'}$. Since only one line can have these properties, $\overleftrightarrow{CC'} = \ell$, and so $\overrightarrow{CC'} \parallel \overleftrightarrow{AA'}$.

Following the approach of Chapter V, let us tentatively define an *ideal point* in space to be the family consisting of a given line and all lines parallel to it in a given direction. Then as before, a unique line is determined by a real point and an ideal point, or by two distinct ideal points (cf. Exercise 1, Section V-10).

Definition VI-16

A line is said to be *parallel* to a plane if it lies in the plane or if it is parallel to its projection onto the plane.

Lemma VI-17

Given two distinct planes, there is a third plane which is perpendicular to both.

Proof. Let α and β be the two given planes. We shall construct a plane γ perpendicular to α and β. On α, choose any point A that is not also on β. Drop the perpendicular $\overline{AB}$ from A to β. If $\overleftrightarrow{AB}$ lies in α, then α ⊥ β, and so for γ we may take any plane perpendicular to α ∩ β (by Theorem VI-6).

If $\overrightarrow{AB}$ does not lie in α, then drop the perpendicular $\overline{BC}$ from B to α. If $\overline{AB}$ and $\overline{BC}$ coincide (and yield a common perpendicular to

PARALLEL LINES AND PLANES

α and β), then we may take γ to be any plane passing through $\overline{AB}$. Otherwise, let γ be plane ABC (Fig. VI-10), which is perpendicular to both α and β, by Theorem VI-6.

The plane whose existence is guaranteed by the preceding lemma is never unique.

Having established the lemma, we can now show that (just as for lines) there are three possibilities for the mutual disposition of two planes: intersection, divergence, and parallelism. Let α and β be any two distinct planes, and let a third plane γ which is perpendicular to both of them intersect them respectively in lines $\overleftrightarrow{AA'}$ and $\overleftrightarrow{BB'}$. There are three cases to consider.

Case 1: $\overleftrightarrow{AA'}$ and $\overleftrightarrow{BB'}$ intersect.

Case 2: $\overleftrightarrow{AA'}$ and $\overleftrightarrow{BB'}$ are divergent. In this case, the common perpendicular to $\overleftrightarrow{AA'}$ and $\overleftrightarrow{BB'}$ is a common perpendicular to α and β (Lemma VI-7). Any plane through this common perpendicular will therefore intersect α and β in a pair of divergent lines, and so in all directions from the common perpendicular, the distance between α and β increases without bound.

Case 3: $\overleftrightarrow{AA'}$ and $\overleftrightarrow{BB'}$ are parallel. (If necessary, relabel the lines so that the direction of parallelism is as indicated by the symbols $\overrightarrow{AA'} \parallel \overrightarrow{BB'}$.) In this case α and β cannot meet, for if they did, then, by Lemma VI-11, γ would have to be perpendicular to their line of intersection, and then $\overleftrightarrow{AA'}$ and $\overleftrightarrow{BB'}$ would intersect on this line (at the unique point common to α, β, and γ—draw a sketch).

Definition VI-18

Two distinct planes are said to be *parallel* if there exists a third plane which is perpendicular to both, and which intersects them in parallel lines.

A few moments' reflection should convince you that the three cases above are mutually exclusive. Accordingly, if α and β are paral-

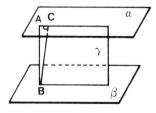

Figure VI - 10.

lel, then *any* plane γ which is perependicular to both will meet them in parallel lines. By the following lemma, the ideal point so determined is independent of γ.

Lemma VI-19

Suppose α and β are parallel planes. Let γ and γ' be planes each of which intersects α and β in a pair of parallel lines. Then all four lines are parallel in the same direction.

Proof. Let Ω be the ideal point common to $\alpha \cap \gamma$ and $\beta \cap \gamma$; let Ω' be the ideal point common to $\alpha \cap \gamma'$ and $\beta \cap \gamma'$. Suppose $\Omega \neq \Omega'$. By Exercise 1 of Section V-10, in each of the planes α and β there is a unique line through Ω and Ω'. But then these two distinct lines are each parallel to each other in both directions. This is impossible. Consequently, Ω and Ω' must coincide.

According to the preceding lemma, two parallel planes have a unique ideal point in common. In view of this, we will amend our tentative definition of ideal point in space, as follows.

Definition VI-20

The family consisting of a given line together with all lines and planes parallel to it in a given direction is called an *ideal point*.

A plane belongs to an ideal point if and only if it contains some line of the ideal point (Exercise 2). Extending our terminology (Section 6, Chapter V), let us agree to say that a line or a plane "passes through" or "contains" an ideal point if it is a member of the set comprising the ideal point. Clearly, if Ω is an ideal point, then two lines passing through Ω, or two points not collinear with Ω determine a unique plane through Ω. Moreover, if two planes through Ω intersect, then their line of intersection passes through Ω (Lemma VI-14).

The following theorem, often thought of as the three-dimensional analog of Playfair's Postulate (Section IV-1), will play a key role in what follows.

Theorem VI-21

Through a given line parallel to a given plane one and only one plane may be drawn parallel to the given plane. All other planes through the given line intersect the given plane.

PARALLEL LINES AND PLANES

Proof. In Figure VI-11, $\overleftrightarrow{AA'}$ is parallel to plane α. Let $\overleftrightarrow{BB'}$ = $\text{proj}_\alpha \overleftrightarrow{AA'}$, where B is the foot of the perpendicular dropped from A to α. Then, by Definitions VI-16 and VI-18, $\overleftrightarrow{AA'} \parallel \overleftrightarrow{BB'}$, and the plane γ which passes through $\overleftrightarrow{AA'}$ and is perependicular to plane AA'BB' is parallel to α. (γ is not shown in Fig. VI-11.)

Suppose δ is any other plane through $\overleftrightarrow{AA'}$ and let $\overleftrightarrow{AC}$ = $\text{proj}_\delta \overleftrightarrow{AB}$, where C is the foot of the perpendicular from B to δ. (Since δ is not perependicular to AA'BB', C is not on $\overleftrightarrow{AA'}$.) By Lemma VI-10, $\sphericalangle BAC < \sphericalangle BAA' = \Pi(AB)$. Consequently, line $\overleftrightarrow{AC}$ — and hence δ — must meet α.

Lemma VI-22

If a plane α does not pass through a given ideal point Ω, then there is a unique line through Ω which is perependicular to α. Moreover, if P is the intersection of this line with α, then the projection onto α of any other line through Ω is contained in an open ray emanating from P.

Proof. Let $Q \in \alpha$, and draw $\overleftrightarrow{Q\Omega}$. If $\overleftrightarrow{Q\Omega} \perp \alpha$, then the first statement in the lemma is immediate. If not, consider the projection $\overleftrightarrow{PQ}$ of $\overleftrightarrow{Q\Omega}$ onto α, where P is chosen so that $\Pi(PQ) = \sphericalangle \Omega QP$ (Fig. VI-12). Then $\overleftrightarrow{P\Omega} \perp \overleftrightarrow{PQ}$ and so $\overleftrightarrow{P\Omega} \perp \alpha$ (Lemma VI-7). Now if $\overleftrightarrow{R\Omega}$ is any other line through Ω, then since plane $RP\Omega \perp \alpha$, the projection of $\overleftrightarrow{R\Omega}$ onto α is contained in line $\alpha \cap RP\Omega$. Of course, $\overleftrightarrow{R\Omega}$ may intersect, be divergent with, or be parallel to its projection on α. As a consequence, $\text{proj}_\alpha \overleftrightarrow{R\Omega}$ will be either an open segment (in the first and second cases) or an open ray (in the third case). (See Exercise 1 of Section 1.) In any event, P is an end point of $\text{proj}_\alpha \overleftrightarrow{R\Omega}$ but is not contained in it.

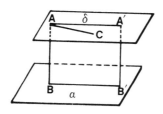

Figure VI - 11.

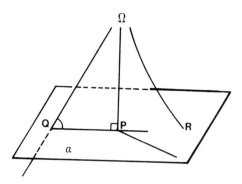

Figure VI - 12.

Exercises VI-2

1. Prove that two lines perpendicular to the same plane are coplanar.

2. Let Ω be the ideal point (Definition VI-20) determined by the line $\overleftrightarrow{AA'}$ (with direction $\overrightarrow{AA'}$). Prove that a plane α belongs to Ω if and only if α contains a line parallel to $\overleftrightarrow{AA'}$ (cf. Definition VI-16).

3. Let $\alpha, \beta,$ and γ be distinct planes. If $\alpha \parallel \beta$ and $\beta \parallel \gamma$, need α be parallel to γ? Explain.

4. (*The Klein-Beltrami Ball*) The 3-dimensional generalization of the Klein-Beltrami Disk is obtained by taking a fixed sphere Σ and adopting the following interpretations.

"point" = interior point of Σ
"line" = open chord of Σ
"plane" = (non-empty) intersection of a plane with the interior of Σ (this is an open disk)

Then, as in the 2-dimensional model, two "lines" are "parallel" if and only if they are chords with a common end point (on Σ).

(a) How would a "plane and a line parallel to it" be represented on this model?

(b) How would two "parallel planes" be represented? What is an "ideal point"?

(c) (See Exercise 2, Section V-10.) Describe "perpendicularity of planes"—or "perpendicularity of a line and a plane"—in this

PARALLEL LINES AND PLANES

model. How would you construct the set of all "lines and planes perpendicular to a given plane"?

(d) Given two "parallel planes," α and β in this model, can you describe the family of "planes" which are "perpendicular to α and β" and meet them in "parallel lines"?

3. THE LIMITING SURFACE

Definition VI-23

Let Ω be an ideal point (in space). Two distinct (real) points P and Q are said to *correspond* relative to Ω if and only if the line joining the midpoint of segment $\overline{PQ}$ to Ω is a perpendicular bisector of the segment. We shall write $P \stackrel{\frown}{=} Q$.

Equivalently, $P \stackrel{\frown}{=} Q$ if and only if segment $\overline{PQ}$ makes equal angles with rays $\overrightarrow{P\Omega}$ and $\overrightarrow{Q\Omega}$.

Theorem VI-24

Let P, Q, and R be any three distinct (real) points, and let Ω be an ideal point. If, relative to Ω, $P \stackrel{\frown}{=} Q$ and $Q \stackrel{\frown}{=} R$, then $P \stackrel{\frown}{=} R$.

Proof. Let α = plane PQR. The case in which Ω lies on α has been dealt with in Chapter V, Section 11. We therefore assume Ω is not in α. Let K and L be the midpoints of segments $\overline{PQ}$ and $\overline{PR}$, respectively, and let O be the unique point of α for which $\overleftrightarrow{O\Omega} \perp \alpha$ (Fig. VI-13). Draw the remaining lines of Figure VI-13. The remaining steps in the proof are these.

   a. $\overleftrightarrow{KO} \supset \text{proj}_\alpha \overleftrightarrow{K\Omega}$, $\overleftrightarrow{LO} \supset \text{proj}_\alpha \overleftrightarrow{L\Omega}$ (Lemma VI-22)
   b. $\overleftrightarrow{K\Omega} \perp \overline{PQ}$ (since $P \stackrel{\frown}{=} Q$)
   c. $\overleftrightarrow{KO}$ is a perpendicular bisector of $\overline{PQ}$ (Lemma VI-12)
   d. OP = OQ
   e. similarly, OQ = OR
   f. OP = OR
   g. $\triangle POL \cong \triangle LOR$ (SSS)
   h. $\overleftrightarrow{OL} \perp \overleftrightarrow{PR}$
   i. $\overleftrightarrow{L\Omega}$ is a perpendicular bisector of $\overline{PR}$ (Lemma VI-12)

This establishes that $P \stackrel{\frown}{=} R$.

If we make the convention that any point corresponds to itself, then correspondence relative to a given ideal point becomes an equivalence relation, as in two dimensions.

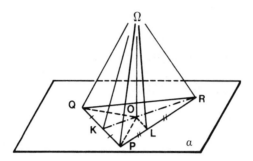

Figure VI - 13.

Definition VI-25

The set of all points which correspond to a given real point relative to a given ideal point Ω is called a *horosphere* or *limiting surface*. Ω is called the *center*, and the lines through Ω are called the *axes* or *radii* of the limiting surface. Planes through Ω are called *diametral planes*. Two horospheres with the same center are called *concentric*.

Since correspondence with respect to Ω is an equivalence relation, a horosphere is uniquely determined by its center and any one of its points (just as an ordinary sphere is determined by its center and one of its points).

If Ω is an ideal point in space, and α is any plane through Ω, then the lines through Ω which lie on α form an ideal point in the planar sense (Definition V-13). Therefore, if σ is a horosphere with center Ω, then any diametral plane intersects σ in a limiting curve (Fig. VI-14). It follows from Theorem V-27 that σ is not a plane.

From Lemma VI-22 it is possible to show that any plane which meets σ, but does not pass through Ω, either intersects σ in a circle or is tangent to σ at one point. It can then be shown that σ is intersected orthogonally by each of its axes. These facts will not be crucial to us, and so the details are omitted.

Notice the similarity with spherical geometry, where a plane through the sphere's center cuts the sphere in a great circle, and a plane which meets the sphere but does not pass through its center either intersects the sphere in a (non-great) circle or is tangent to it. On a horosphere, the counterparts of great circles are limiting curves. (Remember, though, that a limiting curve is not a closed curve, nor is a horosphere a closed surface.)

THE LIMITING SURFACE

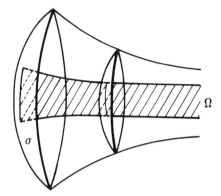

Figure VI - 14.

This similarity can be exploited much further. Spherical geometry is the geometry that holds on a certain surface (a sphere) in Euclidean 3-space, when the primitive geometric terms are suitably interpreted (e.g., "line" means "great circle," a plane section through the sphere's center). Although constructed within Euclidean 3-space, the resulting 2-dimensional geometry is quite non-Euclidean (any two "lines" intersect).

Quite analogously, it is natural to ask what geometry is obtained from a horosphere in Lobachevskian 3-space if we take as the "lines" of this geometry, the horocycles (plane sections through the horosphere's center). What is the nature of the resulting 2-dimensional geometry? Is it non-Lobachevskian?

Horospherical Geometry

To answer the above questions, let us now examine in detail the geometry we obtain on a horosphere σ by interpreting the primitive term "point" to mean point of σ and by interpreting the primitive term "line" to mean a plane section of σ by a diametral plane, i.e., a horocycle on σ. Since a "line" then is a set of "points," the meanings to be assigned to such expressions as "lies on," "passes through," "contains," etc., are obvious. Any two distinct "points" P and Q determine a unique diametral plane, which intersects σ in a horocycle. Therefore, there exists a unique "line" joining two given distinct "points."

In Section V-12, we saw how the concepts of betweenness and congruence could be introduced for limiting arcs. Accordingly, it is easily verified that the geometry of σ ("horospherical geometry") satisfies Hilbert's axioms I,1,2,3, II,1,2,3, and III,1,2,3. (A "segment" in this geometry is a limiting arc.)

Two horocyclic "rays" of σ, emanating from a common "point" P, form an "angle." This "angle" may be measured by the angle between the tangents to these arcs at P. Since these tangents are both perpendicular to the radius through P and lie in the planes of these arcs, then this angle is also the plane angle of the dihedral angle formed by these planes. Hilbert's axiom III-4 for horospherical geometry then follows from this axiom's validity in Lobachevskian 3-space.

Theorem VI-26

The geometry of a horosphere is Euclidean.

Proof. We must check that the postulates of Hilbert, or of an equivalent set of postulates, hold true in the geometry of σ. For the sake of brevity, however, we shall verify only the more interesting postulates and content ourselves with a few remarks on some of the others.

Most of the axioms of incidence, order, and congruence have already been established. To prove Pasch's Axiom, it will suffice to prove the equivalent Separation Axiom (cf. Section III-3).

If ℓ is any "line" on the horosphere with center Ω, then plane $\alpha = \ell\Omega$ separates the points of space not on α into two convex* sets called *half-spaces* (see Exercise 22 of Section III-3). Let H and K be the intersections of these two half-spaces with σ. If $A \in H$ and $B \in K$, then segment $\overline{AB}$ meets α in a point P' (Fig. VI-15). Let $P = \overleftrightarrow{P'\Omega} \cap \sigma$. Then, since P' is between A and B on $\overline{AB}$, P lies on $\widehat{AB}$ (reread the definition of "between" for limiting arcs in Section V-12). Since P is on α as well as on σ, P is on $\ell = \alpha \cap \sigma$. Thus $\widehat{AB}$ intersects σ. Similarly, it can be shown that the "segment" joining two points of H or two points of K does not meet ℓ. Thus, the Separation Axiom holds on σ.

We shall not stop to prove the SAS postulate. It can be deduced from absolute theorems concerning trihedral angles (a concept with

---

*A set is called *convex* if the segment joining any two of its points lies entirely in the set.

THE LIMITING SURFACE

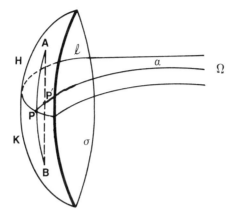

Figure VI - 15.

which you may be unfamiliar). It can be verified also by first showing that σ possesses certain transformations or rigid motions: rotations (about any axis),* translations (along any of its horocycles), and reflections (in any of its diametral planes). The SAS postulate can then be established by moving one triangle onto another, just as in Euclid's original proof (cf. Kulczycki [27, p. 122]).

Continuity (Dedekind's Postulate) was established in Section V-12.

Finally, Playfair's Postulate can be deduced from Theorem VI-21. If P is a "point" of σ which does not lie on a "line" ℓ of σ, then line $\overleftrightarrow{P\Omega}$ is parallel to plane ℓΩ. By Theorem VI-21, there is a unique diametral plane α through PΩ and parallel to ℓΩ, and any other diametral plane through ℓ intersects ℓΩ. Accordingly, σ ∩ α is the unique "line" which passes through P and does not meet ℓ.

Another way of proving the Euclidean parallel postulate is to prove the equivalent statement that through any three points P, Q, and R of σ, not on the same "line," a circle may be drawn. In the course of proving Theorem VI-24, we showed that any three points P, Q, and R of σ are equidistant from a point 0 (which is not on σ), and so lie on a circle in plane PQR. To complete the argument, one would need to show that this circle lies entirely on σ, and that all of

---

*A horosphere is actually a surface of revolution, obtained by revolving a horocycle about any one of its axes.

its points are equidistant from the point $\overleftrightarrow{O\Omega} \cap \sigma$ in the geometry of σ (where distances are measured along limiting arcs) (see Exercise 1).

Now that you have seen a surface in Lobachevskian 3-space whose intrinsic geometry is Euclidean, you may wonder if, dually, there exists a surface in Euclidean 3-space whose intrinsic geometry is Lobachevskian (with a suitable interpretation of the word "line"). In 1868, Eugenio Beltrami, in his *Essay on the interpretation of non-euclidean geometry*, described a surface in Euclidean space—the *pseudosphere*—that gives a partial representation of Lobachevskian geometry, in the sense that it is a model for a portion of the Lobachevskian plane.

To construct the pseudosphere, we start with a plane curve known as the *tractrix* or *drag curve*, so called because it is the curved traced by an object being pulled along the ground at the end of a rope. For example, imagine a boy initially standing at the origin of a Cartesian coordinate system in the Euclidean plane. In his hand is the end of a taut leash attached to a dog located at the point (0,a) on the y axis. The boy begins to walk along the positive x axis (toward the veterinarian's office). The dog resists and is dragged along a curved path (the dotted curve in Fig. VI-16a). As the dog traces this path, the leash is always tangent to the curve at the dog's location. This curve therefore has the property that the distance from the point of contact of one of its tangent lines to the point where that tangent line meet's the curve's asymptote (the x axis) is a constant, a (= the length of the leash).

The surface S obtained by revolving the tractrix about its asymptote is the pseudosphere (Fig. VI-16b). The "lines" of the

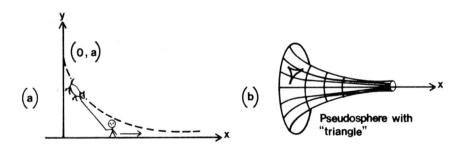

Figure VI - 16.

THE LIMITING SURFACE

pseudosphere are curves known as geodesics. The term *geodesic* will not be defined precisely in this book. Suffice it to say that, roughly speaking, a curve α on S is a geodesic if for any two points A and B on S, α is the (or a) shortest curve in S connecting A to B. (A rubber band, stretched between A and B and constrained to lie in the surface, will (in the absence of friction, contract until it forms a geodesic arc.)

Beltrami showed that the geodesic geometry of the pseudosphere is such that the sum of the angles of any triangle is less than a straight angle, and the triangle's area is proportional to its defect. Euclidean planes through the x axis intersect the pseudosphere in "lines" (actually tractrices) that comprise an ideal point, Ω.

The pseudosphere, unfortunately, represents only part of the Lobachevskian plane, since it contains only one ideal point, Ω, described above. Moreover, "lines" through Ω cannot be extended indefinitely in the opposite direction, since the pseudosphere cannot be smoothly extended beyond its edge. The pseudosphere of Figure VI-16b represents the Lobachevskian region bounded by a horocyclic arc and the two radii through the arc's end points. However, corresponding points of these radii (Definition V-25) are identified. Thus the pseudosphere can be thought of as a piece of the Lobachevskian plane that has been rolled up to form a horn, much as a Euclidean rectangle can be rolled up to form a cylinder. Since planes perpendicular to the x-axis cut the pseudosphere in curves orthogonal to all the "lines" of Ω, these curves represent concentric horocyclic arcs on the pseudosphere, but with end points identified.

Thus, the pseudosphere is only "locally" a model for Lobachevskian geometry. In fact, Hilbert, in 1901, proved that there does not exist in Euclidean 3-space any smooth surface whose geodesic geometry represents the entire Lobachevskian plane. Nevertheless, it was Beltrami's pseudosphere which, more than anything else, convinced mathematicians that Lobachevsky's geometry was as consistent as Euclid's.

We conclude this section with an important fact concerning limiting arcs.

Lemma VI-27

There is a unique length S such that any limiting arc of length S has its tangent line at either end parallel to the radius through the other. For any limiting arc of length less than S, the tangent line at either end intersects the radius through the other.

Proof. Let $\widehat{AB}$ be a horocyclic arc with center Ω. Let C be the foot of the perpendicular from A to the radius $\overleftrightarrow{B\Omega}$, and let $x = AC$. Draw $\overleftrightarrow{AT}$ tangent to $\widehat{AB}$ at A, where T is on the B side of $\overleftrightarrow{A\Omega}$ (Fig. VI-17).

Now $\overleftrightarrow{AT} \parallel \overleftrightarrow{CB}$ if and only if $\angle TAC = \Pi(x)$. But since $\overleftrightarrow{AT} \perp \overleftrightarrow{A\Omega}$, $\angle TAC = \pi/2 - \Pi(x)$. Therefore, $\overleftrightarrow{AT} \parallel \overleftrightarrow{CB}$ if and only if $\Pi(x) = \pi/4$. Accordingly, S is one-half the length of the horocyclic arc subtended by a chord of length 2x, where $\Pi(x) = \pi/4$.

Moreover, $\overrightarrow{AT}$ intersects $\overleftrightarrow{CB}$ if and only if $\angle TAC < \Pi(x)$, i.e., if and only if $\Pi(x) > \pi/4$. Since Π is a decreasing function of x, this means that $\overrightarrow{AT}$ intersects $\overleftrightarrow{CB}$ if and only if $\widehat{AB} < S$.

(Note that S is an example of an absolutely determined length; cf. Section IV-3.)

Exercises VI-3

1. Let P, Q, and R be three distinct points of a horosphere σ, not all on the same "line" (horocycle). From the proof of Theorem VI-24, we know these points lie on a circle in a plane α perpendicular to $\overrightarrow{O\Omega}$, where O is the circle's center.

   (a) Show that if S is any other point of this circle, then S is on σ.

   (b) Let $O' = \overleftrightarrow{O\Omega} \cap \sigma$. Prove that if A and B are any two points of the circle, then $\widehat{O'A} = \widehat{O'B}$.

2. Using properties of limiting arcs and their segments, deduce that Hilbert's axioms III-1, 2 and 3 hold in horospherical geometry. (Assume the results of Section V-12 hold in Lobachevskian 3-space.)

3. *Definition*: An *ultra-ideal point* (in space) is the set of all lines and planes perpendicular to a given plane. This plane is called the *carrier* of the ultra-ideal point.

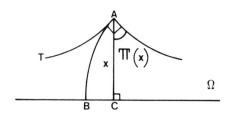

Figure VI - 17.

THE LIMITING SURFACE

Let Γ be the ultra-ideal point with carrier α. Say that two (real) points P and Q *correspond* relative to Γ—indicated by $P \underset{\Gamma}{\simeq} Q$, or just $P \simeq Q$—if the perpendicular to α through the midpoint of $\overline{PQ}$ is a perpendicular bisector of $\overline{PQ}$.

(a) Let P and Q be distinct points. Prove that $P \simeq Q$ if and only if P and Q are equidistant from and on the same side of α. (From this, it then follows that if P, Q, and R are distinct points such that $P \simeq Q$ and $Q \simeq R$, then $P \simeq R$.)

(b) Define an *equidistant surface* and prove it is not a plane.

4. Let P be any point of a horosphere σ with center Ω. Show that the plane through P that is perpendicular to $\overleftrightarrow{P\Omega}$ has no other point in common with σ.

4. THE ANGLE OF PARALLELISM FORMULA

Suppose ϕ is a plane tangent to the horosphere σ with center Ω. For each point P of ϕ, the radius $\overleftrightarrow{P\Omega}$ meets σ in a unique point P' (Fig. VI-18). The association $P \to P'$ furnishes a one-to-one mapping from ϕ into σ (see Exercise 1). Under this mapping, lines in ϕ correspond to limiting arcs on σ. Accordingly, if $\triangle ABC$ is a triangle on ϕ, then A', B', and C' determine a horospherical triangle, which we will denote $\widehat{\triangle} A'B'C'$.

Since, as we showed in the preceding section, the geometry of σ is Euclidean, the formulas of ordinary trigonometry apply to $\widehat{\triangle} A'B'C'$. You might suspect that if we could learn enough about the mapping $P \to P'$, i.e., how it affects distances and angles, then we might relate the Euclidean trigonometric relationships that hold in $\widehat{\triangle} A'B'C'$ to the Lobachevskian relationships that hold in $\triangle ABC$, and so derive the laws of Lobachevskian trigonometry. This is exactly

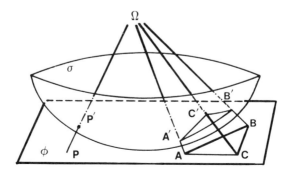

Figure VI-18.

what Lobachevsky did, and we follow his basic idea in the derivations below. Our first task will be to find an explicit formula for the angle of parallelism function, $\Pi(x)$. To this end, we begin by applying the above plan to an ideal triangle in the tangent plane.

Let $\triangle AC\Omega$ be an ideal triangle with right angle at C. Let $x = AC$ and $\alpha = \angle CA\Omega = \Pi(x)$. At A, erect $\overleftrightarrow{A\Omega'}$ perpendicular to the plane of the triangle, and draw the parallels $\overleftrightarrow{C\Omega'}$ and $\overleftrightarrow{\Omega\Omega'}$ (Fig. VI-19). For convenience, let

I = plane $AC\Omega$, II = plane $AC\Omega'$, III = plane $C\Omega\Omega'$

Then

    a. $\angle CA\Omega' = \pi/2$ (Definition VI-1)
    b. $\angle AC\Omega' = \Pi(x) = \alpha$
    c. I $\perp$ II (Theorem VI-6)
    d. $\overleftrightarrow{C\Omega} \perp$ II (Lemma VI-7)
    e. $\angle \Omega C\Omega' = \pi/2$ (Definition VI-1)
    f. II $\perp$ III (Theorem VI-6, since $\overleftrightarrow{C\Omega}$ lies in III, $\overleftrightarrow{C\Omega} \perp$ II)

Now draw the horosphere passing through A and centered at Ω'. This horosphere, which is tangent to I at A, meets $\overleftrightarrow{C\Omega'}$ and $\overleftrightarrow{\Omega\Omega'}$ in points C' and B', respectively, and determines a horospherical triangle, $\widehat{\triangle}AB'C'$ (Fig. VI-20). Since II $\perp$ III, $\angle C'$ of $\widehat{\triangle}AB'C'$ is right (the tangent lines to $\overline{AC'}$ and $\overline{B'C'}$ at C' meet radius $\overleftrightarrow{C\Omega'}$ orthogonally, and form a plane angle of a right dihedral angle). In addition,

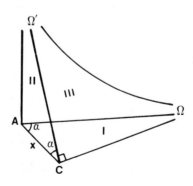

Figure VI - 19.

THE ANGLE OF PARALLELISM FORMULA

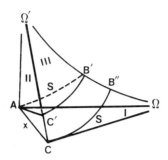

Figure VI - 20.

∡A of $\widehat{\triangle}$ AB'C' is congruent to α (since $\overrightarrow{AC}$ and $\overrightarrow{A\Omega}$ are tangent to its sides), and the length of the hypotenuse, $\widehat{AB'}$, equals S, by Lemma VI-27.

By Euclidean trigonometry, we have $\widehat{B'C'} = S \sin \alpha$ and

$$\widehat{AC'} = S \cos \alpha \tag{11}$$

Draw the limiting arc $\widehat{B''C}$ concentric with $\widehat{B'C'}$ (with B" on $\overleftrightarrow{\Omega\Omega'}$). Since $\overleftrightarrow{\Omega\Omega'}$ is parallel to both sides of right angle ∡ΩCΩ', $\widehat{B''C} = S$. By Theorem V-34, $e^{CC'/k} = S/\widehat{B'C'} = S/(S \sin \alpha)$ or

$$e^{CC'/k} = 1/\sin \alpha \tag{12}$$

Next, in plane II, extend $\overrightarrow{CA}$ through A to ideal point Ω'' and draw $\overrightarrow{\Omega'\Omega''}$ (Fig. VI-21). On ray $\overrightarrow{C\Omega'}$, mark off CD = AC = x and

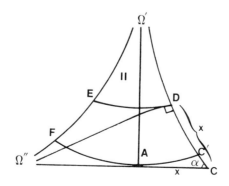

Figure VI - 21.

draw $\overleftrightarrow{D\Omega''}$. Since $\alpha = \Pi(x)$, $\overleftrightarrow{D\Omega''} \perp \overleftrightarrow{C\Omega'}$. Extend limiting arc $\widehat{C'A}$ through A to F on $\overleftrightarrow{\Omega'\Omega''}$, and draw through D the concentric limiting arc $\widehat{DE}$ meeting $\overleftrightarrow{\Omega'\Omega''}$ in E. Then $\widehat{DE} = \widehat{AF} = S$. By Theorem V-34 again, $e^{C'D/k} = \widehat{C'F}/\widehat{DE} = (\widehat{AC'} + S)/S = (S \cos \alpha + S)/S$, or

$$e^{C'D/k} = 1 + \cos \alpha \tag{13}$$

Multiplying (12) and (13), and adding exponents, we obtain

$$\frac{1 + \cos \alpha}{\sin \alpha} = e^{x/k} \tag{14}$$

where $\alpha = \Pi(x)$. Using the identities

$$\cos \alpha = \cos\left(\frac{\alpha}{2} + \frac{\alpha}{2}\right) = \cos^2 \frac{\alpha}{2} - \sin^2 \frac{\alpha}{2} = 2\cos^2 \frac{\alpha}{2} - 1$$

$$\sin \alpha = 2 \sin \frac{\alpha}{2} \cos \frac{\alpha}{2}$$

we may rewrite (14) in the equivalent form

$$\boxed{\tan \frac{\Pi(x)}{2} = e^{-x/k}} \tag{15}$$

the angle of parallelism formula derived independently by both Bolyai and Lobachevsky.

Theorem VI-28

$\Pi(x)$ equals twice the angle whose tangent is $e^{-x/k}$.

From (15) we can obtain formulas for $\cos\Pi(x)$ and $\sin\Pi(x)$:

$$\cos \alpha = \frac{\cos^2 \frac{\alpha}{2} - \sin^2 \frac{\alpha}{2}}{\cos^2 \frac{\alpha}{2} + \sin^2 \frac{\alpha}{2}} = \frac{1 - \tan^2 \frac{\alpha}{2}}{1 + \tan^2 \frac{\alpha}{2}} = \frac{1 - e^{-2x/k}}{1 + e^{-2x/k}} = \tanh(x/k)$$

[In the next to the last expression, multiply numerator and denominator by $e^{x/k}$ to obtain $\tanh(x/k)$.] Since α is acute, we take the positive square root of $(1 - \cos^2 \alpha)$ to obtain $\sin \alpha = [1 - \tanh^2(x/k)]^{1/2} = \text{sech}(x/k)$. Therefore,

THE ANGLE OF PARALLELISM FORMULA

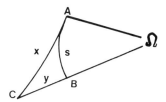

Figure VI - 22.

$$\cos \Pi(x) = \tanh(x/k) \qquad (16)$$

$$\sin \Pi(x) = \text{sech}(x/k) \qquad (17)$$

Using (16) and (17), we can now rewrite (11) and (12) as $\widehat{AC'} = S \tanh(x/k)$, $e^{CC'/k} = \cosh(x/k)$, and so, as a by-product of our calculations, we have the following lemma, which will be needed in the next section.

Lemma VI-29

Let $\widehat{AB}$ be a limiting arc of length $s < S$, and let C be the intersection of the tangent line at A with the radius through B (Fig. VI-22). If $x = AC$ and $y = BC$, then

$$s = S \tanh(x/k) \qquad (18)$$

$$e^{y/k} = \cosh(x/k) \qquad (19)$$

Exercises VI-4

1. Show that the mapping from the tangent plane ϕ to the horosphere σ, defined by $P \to P' = \overleftrightarrow{P\Omega} \cap \sigma$ is not onto. What is the image of this mapping?

2. Show that the distance x satisfying $\Pi(x) = \pi/4$ can be described as follows. Let $\angle PQR$ be any right angle, and let ℓ be the unique line parallel to both its sides, by Corollary V-10. If T is the foot of the perpendicular dropped from Q to ℓ, then $x = QT$. (x is an example of an absolutely determined length.)

3. Show that $\Pi(x) = \pi/4$ if and only if $x = k \ln(1 + \sqrt{2})$.

4. Let $\widehat{AB}$ be any limiting arc. Let D be the foot of the perpendicular dropped from A to the radius through B. Show that $\widehat{AB} = S \sinh(AD/k)$.

5. Show that the length s of a horocyclic arc subtending a chord of length x is given by $x = 2S \sinh(x/2k)$ (see Exercise 4).

6. In (14), replace $\sin\alpha$ by $(1 - \cos^2\alpha)^{1/2}$ and deduce (16) directly from the resulting equation.

7. Let $y = \Pi(x)$. Then $\tan y/2 = e^{-x/k}$. (a) Show that $dy/dx = -(1/k) \text{sech}(x/k)$. (b) Sketch the graph of $\Pi(x)$.

8. By choosing one convenient limiting arc to serve as a unit, limiting arcs can be assigned numerical measures (lengths). For example, S could be chosen as the unit, and the length of any arc s would then be the ratio s/S. However, it might be more appropriate to choose the unit of arc length to be compatible with the unit of linear distance. This is done by requiring that a small limiting arc s and its chord x have approximately the same measure, with equality approached in the limit as s and x go to zero, i.e.,

$$\lim_{x \to 0} \frac{s}{x} = 1$$

Prove that under such a choice of limiting arc measure, $S = k$ (cf. Exercise 5).

5. TRIANGLE RELATIONS

We may now proceed to find the relationships between the various parts of a right triangle in Lobachevskian geometry. For this purpose, we use a figure similar to Fig. VI-19, except that the ideal vertex Ω will be replaced by a real vertex B (Fig. VI-23). As before, plane II is perpendicular to both planes I and III, and horospherical triangle $\widehat{\triangle} AB'C'$ is right with angle α at vertex A. In $\triangle ABC$, let a, b, and c denote the sides opposite A, B, and C, respectively. The corresponding sides of the horospherical triangle, $\widehat{\triangle} AB'C'$, are denoted a', b', and c'. Let $\beta = \angle ABC$.

Since (in Fig. VI-23) $BB' = BB'' + B''B'$, we have $e^{BB'/k} = e^{BB''/k} e^{B''B'/k}$. However, by Lemma VI-29, $e^{BB'/k} = \cosh(c/k)$, $e^{BB''/k} = \cosh(a/k)$, and $e^{B''B'/k} = e^{CC'/k} = \cosh(b/k)$. There then follows

$$\boxed{\cosh(c/k) = \cosh(a/k) \cosh(b/k)} \tag{20}$$

which is the Lobachevskian analog of the Pythagorean Theorem.

TRIANGLE RELATIONS

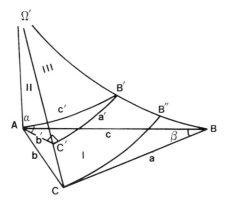

Figure VI - 23.

Lemma VI-29 gives us also $b' = S \tanh(b/k)$, $c' = S \tanh(c/k)$, and $a' = (\widehat{B''C})e^{-CC'/k} = [S \tanh(a/k)]/[\cosh(b/k)]$. Substituting the first two into the Euclidean formula $b' = c' \cos\alpha$, we obtain the Lobachevskian formula

$$\boxed{\tanh(b/k) = \tanh(c/k)\cos\alpha} \tag{21a}$$

(Luckily, the constant S drops out.) The twin formula

$$\boxed{\tanh(a/k) = \tanh(c/k)\cos\beta} \tag{21b}$$

is obtained (not from Fig. VI-23 but) simply by interchanging b and a and interchanging α and β.

From $a' = c' \sin\alpha$, we have

$$\frac{S \tanh(a/k)}{\cosh(b/k)} = S \tanh(c/k) \sin\alpha$$

or

$$\frac{\sinh(a/k)}{\cosh(a/k)\cosh(b/k)} = \frac{\sinh(c/k)}{\cosh(c/k)} \sin\alpha$$

In view of (20), we conclude that

$$\boxed{\sinh(a/k) = \sinh(c/k)\sin\alpha} \qquad (22a)$$

and dually (again interchanging a and b, α and β),

$$\boxed{\sinh(b/k) = \sinh(c/k)\sin\beta} \qquad (22b)$$

From $a' = b'\tan\alpha$, you can easily deduce

$$\boxed{\tanh(a/k) = \sinh(b/k)\tan\alpha} \qquad (23a)$$

and its twin

$$\boxed{\tanh(b/k) = \sinh(a/k)\tan\beta} \qquad (23b)$$

and by multiplying these together and utilizing (20),

$$\boxed{\cosh(c/k) = \cot\alpha\cot\beta} \qquad (24)$$

Now let $\triangle ABC$ be an arbitrary (non-right) triangle with a, b, and c the sides opposite vertices A, B, and C, respectively. We shall use the letters A, B, and C also to denote the angles at the respective vertices. Let h be the length of the perpendicular drawn from vertex B to point D on the line containing the opposite side. Let d = CD. There are three cases to consider depending on the location of D (see Fig. VI-24).

Assuming the order is (CDA) or (CAD), in $\triangle ABD$

$$\cosh\frac{c}{k} = \cosh\frac{h}{k}\cosh\frac{b-d}{k}$$
$$= \cosh\frac{h}{k}\cosh\frac{b}{k}\cosh\frac{d}{k} - \sinh\frac{b}{k}\sinh\frac{d}{k}\cosh\frac{h}{k}$$

Using the Pythagorean relation $\cosh(h/k)\cosh(d/k) = \cosh(a/k)$, and

TRIANGLE RELATIONS

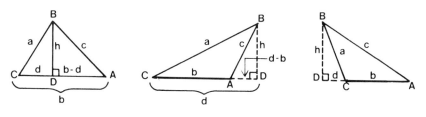

Figure VI - 24.

rewriting $\sinh(d/k)$ as $\tanh(d/k)\cosh(d/k)$, we have

$$\cosh\frac{c}{k} = \cosh\frac{a}{k}\cosh\frac{b}{k} - \sinh\frac{b}{k}\cosh\frac{a}{k}\tanh\frac{d}{k}$$

By (21), $\tanh(d/k) = \tanh(a/k)\cos C$. Therefore we have

$$\boxed{\cosh\frac{c}{k} = \cosh\frac{a}{k}\cosh\frac{b}{k} - \sinh\frac{a}{k}\sinh\frac{b}{k}\cos C} \qquad (25)$$

which is the Lobachevskian Law of Cosines. (The analogous Euclidean formula is $c^2 = a^2 + b^2 - 2ab\cos C$.)

If the ordering is (DCA), one starts with $\cosh(c/k) = \cosh(h/k)\cosh[(b+d)/k]$, and applies (21) to $\pi - \angle C$. As you can check, we obtain the same formula, (25), as before.

Applying (22) to Figure VI-24, we obtain $\sinh(h/k) = \sinh(c/k)\sin A = \sinh(a/k)\sin C$, and so $\sinh(c/k)(\sin C)^{-1} = \sinh(a/k)(\sin A)^{-1}$. [For the orderings (CAD), (DCA), use the identity $\sin(\pi - \theta) = \sin\theta$.] By symmetry, $\sinh(a/k)(\sin A)^{-1} = \sinh(b/k)(\sin B)^{-1}$, and so we have derived the Lobachevskian Law of Sines, for an arbitrary triangle, $\triangle ABC$:

$$\boxed{\frac{\sinh(a/k)}{\sin A} = \frac{\sinh(b/k)}{\sin B} = \frac{\sinh(c/k)}{\sin C}} \qquad (26)$$

(The analogous Euclidean formula is

$$\frac{a}{\sin A} = \frac{b}{\sin B} = \frac{c}{\sin C}\ )$$

Exercises VI-5

In the first four problems, assume that $\triangle ABC$ is a right triangle with right angle at C and that the sides and angles are labeled a, b, c, α, and β as in Figure VI-23. All figures in this section are in the Lobachevskian plane.

1. Show that $\cos \beta = \cosh(b/k) \sin \alpha$.
2. Is there a Euclidean counterpart to (24)? Why?
3. (Use a calculator or tables.) (a) If $a = 3k$ and $b = 4k$, find c (in terms of k). (b) If $a = b = k$, what is c? (c) What would be the corresponding values for c above in Euclidean geometry? (Take k to be any fixed length.)
4. Derive the formula

$$\sin \mathscr{A}(ABC) = \frac{\sinh \frac{a}{k} \sinh \frac{b}{k}}{\cosh \frac{c}{k} + 1}$$

{Hint: $\sin \mathscr{A}(ABC) = \sin[(\pi/2 - \alpha) - \beta] = \cos \alpha \cos \beta - \sin \alpha \sin \beta$.}

5. In the figure, M is the midpoint of $\overline{AB}$. Show that

$$\cosh \frac{m}{k} = \frac{\cosh \frac{a}{k} + \cosh \frac{b}{k}}{2 \cosh \frac{c}{2k}}$$

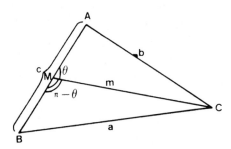

(In Euclidean geometry, the corresponding result is $2m^2 = a^2 + b^2 - c^2/2$.)

6. Assume PRSQ is a Lambert quadrilateral with acute angle at S. By Theorem V-21 ($\overleftrightarrow{PQ}$ and $\overleftrightarrow{RS}$ are diverent), $PR < QS$. This im-

TRIANGLE RELATIONS

plies that a circle centered at P with radius QS will meet $\overline{RS}$ in some point T. Let b = QS = PT. Draw $\overline{PS}$. Let $\alpha = \angle TPQ$, $\beta = \angle RPS$, a = PR, and c = PQ. Applying the formulas of this section to the various right triangles in the figure, we have

$$\sin \alpha = \cos\left(\frac{\pi}{2} - \alpha\right) = \frac{\tanh \frac{a}{k}}{\tanh \frac{b}{k}}$$

$$\cos \beta = \frac{\tanh \frac{a}{k}}{\tanh \frac{PS}{k}}$$

$$\cos \beta = \sin\left(\frac{\pi}{2} - \beta\right) = \frac{\sinh \frac{b}{k}}{\sinh \frac{PS}{k}}$$

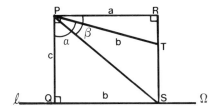

From the above, deduce that $\sin \alpha = \text{sech}(c/k)$. Note: since by (17), $\sin \Pi(c) = \text{sech}(c/k)$, α must equal $\Pi(c)$, and so $\overleftrightarrow{PT} \parallel \overleftrightarrow{QS}$. Consequently, this exercise establishes the validity of the following construction of a line parallel to a given line ℓ through a given external point P.

> Let Ω be the chosen ideal point on ℓ. Drop $\overline{PQ} \perp \ell$. Through P draw $\overleftrightarrow{PR} \perp \overline{PQ}$. From any point S on $\overleftrightarrow{Q\Omega}$, drop $\overline{SR} \perp \overleftrightarrow{PR}$. With P as center and QS as radius, draw an arc meeting $\overline{RS}$ in T. Then $\overrightarrow{PT} \parallel \overrightarrow{Q\Omega}$. (This is J. Bolyai's "straightedge and compass" construction of parallels.)

7. (a) From (16), show that if $\Pi(x) = \pi/3$, then $x = (k/2) \ln 3$.
(b) Prove: in Lobachevskian geometry, a circle of radius at least $(k/2) \ln 3$ cannot be the inscribed circle of any triangle (with real vertices).

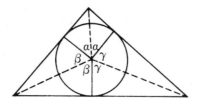

(Hint: suppose such a circle could be inscribed in a triangle, as in the figure. Let r = the radius. Then $r \geq \frac{k}{2} \ln 3$, and so $\Pi(r) \leq \Pi(\frac{k}{2} \ln 3) = \pi/3$, by (a), and because Π is a decreasing function. Show at least one of the three angles α, β, γ must be $\geq \pi/3$ and deduce a contradiction.

8. In the figure, ABCD is a Saccheri quadrilateral with base $\overline{CD}$ and summit $\overline{AB}$.

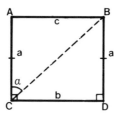

Show that

$$\cosh \frac{c}{k} = \cosh^2 \frac{a}{k} \cosh \frac{b}{k} - \sinh^2 \frac{a}{k}$$

(Hint: draw $\overline{BC}$.)

9. In Exercises IV-5, we learned that area in Lobachevskian geometry is proportional to defect: $A = \alpha \mathscr{D}$ (α constant). According to Exercise 4 then, the area of a right triangle with legs a and b and hypotenuse c is given by

$$A = \alpha \sin^{-1} \left(\frac{\sinh \frac{a}{k} \sinh \frac{b}{k}}{1 + \cosh \frac{c}{k}} \right)$$

Show that the requirement $A \to ab/2$ as $a/k, b/k, c/k \to 0$ implies that $\alpha = k^2$. (See the Note following Exercise 8 of Section IV-5.)

6. THE THREE GEOMETRIES

How are the trigonometric formulas of Euclidean, Lobachevskian, and spherical geometry related? To answer this question, we need to know something about the series representations of both the circular functions (sin, cos, tan, etc.) and the hyperbolic functions (sinh, cosh, tanh, etc.). These series are given in Appendix C. The key fact we need first is this. For $|x|$ sufficiently small compared to unity—we indicate this by writing $|x| \ll 1$—the higher power terms in expansions such as

$$\sinh x = x + \frac{x^3}{3!} + \frac{x^5}{5!} + \frac{x^7}{7!} + \cdots$$

or

$$\tan x = x + \frac{x^3}{3} + \frac{2x^5}{15} + \cdots$$

are negligible compared to the leading terms, and it can be shown that the sum of these higher order terms (let us say, for example, cubed and higher power terms) is likewise negligible compared to the leading terms. In brief, for $|x| \ll 1$,

$$\begin{aligned}\sin x &\approx x \\ \cos x &\approx 1 - x^2/2 \\ \tan x &\approx x; \end{aligned} \qquad (27a)$$

$$\begin{aligned}\sinh x &\approx x \\ \cosh x &\approx 1 + x^2/2 \\ \tanh x &\approx x \end{aligned} \qquad (27b)$$

with the approximations improving as $x \to 0$. (From Taylor's Theorem from advanced calculus, it is known that omitting third and higher order terms results in an error that is less than some constant times $|x|^3$.) Consequently,

$$\begin{aligned}\lim_{x \to 0} \frac{\sin x}{x} &= \lim_{x \to 0} \frac{\tan x}{x} = 1 \\ \lim_{x \to 0} \frac{\sinh x}{x} &= \lim_{x \to 0} \frac{\tanh x}{x} = 1\end{aligned} \qquad (28)$$

Suppose now $\triangle ABC$ is a right triangle in the Lobachevskian plane with $\angle C = \pi/2$. Assume the sides and angles are labeled a, b, c, α, and β as in Figure VI-23. If a, b, and c are sufficiently small compared to the space constant k, then we may use the approximations in (27b) to show that the Lobachevskian triangle relations for $\triangle ABC$ are in close agreement with their Euclidean counterparts.

For example, since $\sinh(a/k) \approx a/k$ and $\sinh(c/k) \approx c/k$, formula (22a) gives, upon multiplication by k,

$$a \approx c \sin \alpha$$

Similarly, because $\tanh(a/k) \approx a/k$, (23a) reduces to

$$a \approx b \tan \alpha$$

Using the approximation $\cosh x \approx 1 + x^2/2$ for $x \ll 1$, we obtain from (20)

$$1 + \frac{c^2}{k^2} \approx (1 + \frac{a^2}{k^2})(1 + \frac{b^2}{k^2}) = 1 + \frac{a^2 + b^2}{k^2} + \frac{a^2 b^2}{k^4}$$

or

$$c^2 \approx a^2 + b^2 + a^2 (\frac{b^2}{k^2}) \approx a^2 + b^2$$

since $b^2/k^2 \ll 1$. Thus a "small" right triangle "almost satisfies" the Pythagorean formula $c^2 = a^2 + b^2$.

Similar analyses can be carried out for the other formulas of Section 5. In general, Euclidean trigonometry gives good approximations in any region whose dimensions are small compared to k. Accordingly, even though empirical observations seem to confirm that Euclidean trigonometry is valid for physical space, we cannot conclude from this that the geometry of the universe is Euclidean: space could conceivably be governed by Lobachevskian geometry, but with k so exceedingly large that only measurements on a galactic scale would reveal a departure from Euclidean geometry.

In an analogous way, using the approximations (27a), we can show that the trigonometric relations that hold for spherical triangles

THE THREE GEOMETRIES

Table VI - 1 Right Triangle Formulas

Euclidean	Lobachevskian	Spherical
$a = c \sin \alpha$	$\sinh \frac{a}{k} = \sinh \frac{c}{k} \sin \alpha$	$\sin \frac{a}{R} = \sin \frac{c}{R} \sin \alpha$
$b = c \sin \beta$	$\sinh \frac{b}{k} = \sinh \frac{c}{k} \sin \beta$	$\sin \frac{b}{R} = \sin \frac{c}{R} \sin \beta$
$b = c \cos \alpha$	$\tanh \frac{b}{k} = \tanh \frac{c}{k} \cos \alpha$	$\tan \frac{b}{R} = \tan \frac{c}{R} \cos \alpha$
$a = c \cos \beta$	$\tanh \frac{a}{k} = \tanh \frac{c}{k} \cos \beta$	$\tan \frac{a}{R} = \tan \frac{c}{R} \cos \beta$
$a = b \tan \alpha$	$\tanh \frac{a}{k} = \sinh \frac{b}{k} \tan \alpha$	$\tan \frac{a}{R} = \sin \frac{b}{R} \tan \alpha$
$b = a \tan \beta$	$\tanh \frac{b}{k} = \sinh \frac{a}{k} \tan \beta$	$\tan \frac{b}{R} = \sin \frac{a}{R} \tan \beta$
$c^2 = a^2 + b^2$	$\cosh \frac{c}{k} = \cosh \frac{a}{k} \cosh \frac{b}{k}$	$\cos \frac{c}{R} = \cos \frac{a}{R} \cos \frac{b}{R}$

(see Table VI-1, third column) are also well approximated by the corresponding Euclidean formulas,* provided the sides of the figures are small compared to R, the sphere's radius. In this case, the closeness of the approximation is believable on intuitive grounds, since we know that a small region of a huge sphere (e.g., a football field on the earth) is practically indistinguishable from a portion of a flat plane. (In fact, a sufficiently small portion of any smooth surface—e.g., a pseudosphere, as well as a sphere—is approximately Euclidean.)

The close similarity in form between spherical formulas and their Lobachevskian counterparts is related to a connection between the circular and hyperbolic functions. These functions, as well as the exponential function, e^x, can be extended to functions of a complex variable by taking their power series (with the variable real or complex) as definitions. For example, sin z, for z complex, is simply defined to be the sum of the convergent series

$$z - \frac{z^3}{3!} + \frac{z^5}{5!} + \frac{z^7}{7!} + \cdots$$

---

*The formulas of spherical trigonometry, with which you may be unfamiliar, are not derived in this text. Take them on faith here.

If $i = (-1)^{1/2}$, then $i^2 = -1$, $i^3 = -i$, $i^4 = 1$, and in general, $i^{n+4} = i^n$, for any integer n. Replacing x by $x/i = -ix$ in the power series for sinh(x), we have

$$\sinh \frac{x}{i} = (-ix) + \frac{(-ix)^3}{3!} + \frac{(-ix)^5}{5!} + \cdots$$

$$= -ix + i\frac{x^3}{3!} - i\frac{x^5}{5!} + \cdots$$

$$= -i(x - \frac{x^3}{3!} + \frac{x^5}{5!} - \cdots) = -i \sin x$$

and so the identity

$$\sinh \frac{x}{i} = -i \sin x \tag{29a}$$

(valid for all x, real or complex). In a similar fashion, we can derive

$$\cosh \frac{x}{i} = \cos x \tag{29b}$$

$$\tanh \frac{x}{i} = -i \tan x \tag{29c}$$

If we now just formally replace k by iR in any of the formulas of Lobachevskian trigonometry (Table VI-1), a remarkable thing happens: we obtain the corresponding formula from spherical trigonometry! For example, (23a) gives, via (29),

$$-i \tan \frac{a}{R} = \tanh \frac{a}{iR} = \sinh \frac{b}{iR} \tan \alpha = -i \sin \frac{b}{R} \tan \alpha$$

or, canceling $-i$,

$$\tan \frac{a}{R} = \sin \frac{b}{R} \tan \alpha$$

a formula of spherical trigonometry. In like manner, the Lobachevskian Law of Cosines, (25), is transformed into

THE THREE GEOMETRIES

$$\cos\frac{c}{R} = \cos\frac{a}{R}\cos\frac{b}{R} - (i\sin\frac{a}{R})(i\sin\frac{b}{R})\cos C$$

$$= \cos\frac{a}{R}\cos\frac{b}{R} + \sin\frac{a}{R}\sin\frac{b}{R}\cos C$$

which is the spherical Law of Cosines.

We can reverse the process as well. Replacing x by x/i in (29) yields the dual relations

$$\sin\frac{x}{i} = -i\sinh x \tag{30a}$$

$$\cos\frac{x}{i} = \cosh x \tag{30b}$$

$$\tan\frac{x}{i} = -i\tanh x \tag{30c}$$

From these, we can verify that replacing R by ik will convert any formula of spherical trigonometry into its Lobachevskian counterpart. For example,

$$\sin\frac{a}{R} = \sin\frac{c}{R}\sin\alpha$$

becomes

$$-i\sinh\frac{a}{k} = -i\sinh\frac{c}{k}\sin\alpha$$

which gives (22a) when $-i$ is canceled.

Lambert's remark that the geometry of the acute angle hypothesis (Section IV-5) would hold on a "sphere of imaginary radius" seems less mysterious now. We can also understand why Lobachevsky was certain his "pangeometry" must be free of contradiction. It seemed unlikely that any inconsistency would not show up in the relationships among the three geometries that we considered in this section.

Exercises VI-6

1. Verify that (25) approximates the Euclidean Law of Cosines when a, b, and c are small compared to k.

2. Show that (23a) approximates the corresponding Euclidean formula (which is what?) when a, b, and c are small compared to k.

3. If $c/k \ll 1$ in (24), what can you conclude about α and β?

4. It is known that, in Lobachevskian geometry, the area A of a right triangle with legs a and b satisfies the relation

$$\tan \frac{A}{2k^2} = \tanh \frac{a}{2k} \tanh \frac{b}{2k}$$

(a) Show that as a, b $\to \infty$, the area A does not $\to \infty$, but approaches a finite limit. What is this limit? (b) Show that for a, b small compared to k, the formula above approximates the Euclidean formula $A = \frac{1}{2}ab$.

5. The area A and circumference C of a circle of radius r in the Lobachevskian plane are

$$A = 4\pi k^2 \sinh^2 \frac{r}{2k} \qquad C = 2\pi k \sinh \frac{r}{k}$$

(see Exercises 7 and 8, below). Show that for $r/k \ll 1$, $A \approx \pi r^2$ and $C \approx 2\pi r$.

6. State the spherical Law of Cosines and the spherical Law of Sines.

7. In the figure, a regular n-gon of side 2x is inscribed in a circle of radius R. By (22a),

$$(*) \quad \sinh \frac{x}{k} = \sinh \frac{R}{k} \sin \frac{\pi}{n}$$

The circumference C_n of this n-gon is 2nx. As $n \to \infty$ (and $x \to 0$), C_n approaches the circumference C of the circle. Use this fact to show that

$$C = 2\pi k \sinh \frac{R}{k}$$

THE THREE GEOMETRIES 245

[Hint: multiply (∗) by 2kn and take the limit, using (28).]

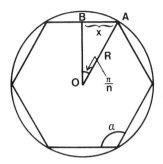

8. In Lobachevskian geometry, the area of any polygon is the product of k^2 and the polygon's defect (cf. Exercises IV-5). Therefore, in the figure of Exercise 7, the area A_n of the regular n-gon inscribed in the circle is

$$A_n = k^2 [2n \; \mathscr{S}(OAB)]$$

Let α be the n-gon's interior angle (twice $\angle BAO$ in the figure in Exercise 7). Note that as $n \to \infty$, $\alpha \to \pi$.

(a) Show $A_n = k^2 [n(\pi - \alpha) - 2\pi]$. (b) Show that for large n, $(\pi - \alpha)/2 \approx [\tanh(x/k)]/\tanh(R/k)$. (c) Using (b) and the approximation $\tanh(x/k) \approx x/k$ for $x/k \ll 1$, show that

$$A_n \approx k^2 \left(\frac{C_n}{k \tanh(R/k)} - 2\pi \right)$$

for n large, where C_n is the n-gon's circumference (see Exercise 7). (d) Deduce that the area of the circle is

$$A = \lim_{n \to \infty} A_n = 4\pi k^2 \sinh^2 \frac{R}{2k}$$

[Hint: $\cosh u - 1 = 2 \sinh^2 (u/2)$, for all u.]

9. In this problem you will find the area of the unbounded region formed by a limiting arc $\widehat{AB}$ of length k and the radii through its end points, i.e., area (ABΩ) in Figure A on the next page.

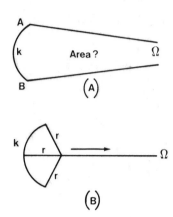

Since a limiting curve is the limit of a circle as the radius r → ∞ (see Exercise 7 of Section V-12), we shall do this by computing the area of a circular sector of radius r determined by a circular arc of length k (Figure B), and then lettering r → ∞. The area A and circumference C of an entire circle of radius r in the Lobachevskian plane are (from Exercises 7 and 8)

$$A = 4\pi k^2 \sinh^2 \frac{r}{2k}, \quad C = 2\pi k \sinh \frac{r}{k}$$

The circular sector in Figure B is only a fraction of a complete circular region. In fact, it is k/C times a whole circle. Using the above expressions for A and C, write an expression for the sector's area and find the limit as r → ∞. [You may use the identity sinh x = 2 sinh(x/2) cosh(x/2) and properties of tanh.]

VII

THE WEIERSTRASS MODEL

> *It is impossible not to feel stirred at the thought of the emotions of men at certain historic moments of adventure and discovery—Columbus when he first saw the Western shore, Pizarro when he stared at the Pacific Ocean, Franklin when the electric spark came from the string of his kite, Galileo when he first turned his telescope to the heavens. Such moments are also granted to students in the abstract regions of thought, and high among them must be placed the morning when Descartes lay in bed and invented the method of co-ordinate geometry.*
>
> —Alfred North Whitehead

INTRODUCTION

In this chapter, we shall demonstrate the consistency of Lobachevskian geometry by constructing a model for it on a certain surface in Euclidean 3-space. This model is probably less familiar than Poincare's disc and upper half-plane models, but is appealing for several reasons. First, it has many properties in common with the sphere as a model for elliptic geometry. For example, "lines" in this model, like great circles, will be intersections of the surface with planes through the origin. In fact, circles, horocycles, and equidistant curves will all turn out to be plane sections. Moreover, this model will introduce the reader to projective concepts (e.g., homogeneous coordinates). It readily generalizes to n dimensions, and it introduces an analogue of the Lorentz metric, central to relativity theory.

1. PRELIMINARIES

Throughout this chapter, we shall denote points in real 3-dimensional coordinate space ($\mathbb{R}^3$) by triples, such as $X = (x_0, x_1, x_2)$. We may think of X not only as a point, but as a *vector*, vizualized as an arrow drawn from the origin, $O = (0, 0, 0)$, to X. Any other arrow obtainable from this one by parallel translation will be considered as representing the same vector. We shall use the terms point and vector interchangeably (see Appendix D).

If $X = (x_0, x_1, x_2)$ and $Y = (y_0, y_1, y_2)$ are any two vectors, $X \cdot Y$ denotes their Euclidean *inner* (or *dot*) *product*:

$$X \cdot Y = x_0 y_0 + x_1 y_1 + x_2 y_2$$

$X \cdot X$ is the square of the distance from O to X. Recall that the angle θ between two non-zero vectors X and Y (pictured as arrows emanating from the same point) is given by

$$\cos \theta = \frac{X \cdot Y}{\sqrt{(X \cdot X)(Y \cdot Y)}} \tag{31}$$

from which it follows that X and Y are *orthogonal* (perpendicular) if and only if $X \cdot Y = 0$. Since $|\cos \theta| \leq 1$, the Cauchy-Schwartz inequality, $(X \cdot Y)^2 \leq (X \cdot X)(Y \cdot Y)$, is immediate from (31).

The equation of a plane in $\mathbb{R}^3$ is a linear equation of the form $a_0 x_0 + a_1 x_1 + a_2 x_2 = b$, or, in vector notation, $A \cdot X = b$. The coefficient vector, $A = (a_0, a_1, a_2)$, is normal to the plane (see Exercise 1).

In the development of a model for Lobachevskian geometry, we shall use another inner product also. The *hyperbolic inner product* of two vectors X and Y is defined by the formula

$$\langle X, Y \rangle = - x_0 y_0 + x_1 y_1 + x_2 y_2$$

For any vector $X = (x_0, x_1, x_2)$, let X_{ref} denote the reflection of X in the $x_1 x_2$-plane:

$$X_{ref} = (-x_0, x_1, x_2)$$

PRELIMINARIES 249

Then the identity

$$<X,Y> = X \cdot Y_{ref} = X_{ref} \cdot Y$$

relates the hyperbolic and Euclidean inner products. From now on, let us say that two vectors X and Y are *h-orthogonal* if $<X,Y> = 0$, and *e-orthogonal* if $X \cdot Y = 0$. Clearly X and Y are h-orthogonal if and only if X_{ref} and Y are e-orthogonal (Fig. VII-1a). Consequently, if the equation of a plane in $\mathbb{R}^3$ is $<A,X> = b$, or equivalently, $A_{ref} \cdot X = b$, then an *h-normal* of the plane, A, can be obtained by first drawing the *e-normal* A_{ref}, and then reflecting in the $x_1 x_2$-plane. Remembering this will help you to draw sketches illustrating theorems that follow.

Note that it is possible for a non-zero vector to be h-orthogonal to itself. In fact, the set of all X satisfying $<X,X> = 0$ forms a cone of revolution with the origin as vertex and semi-vertex angle of 45°.

Figure VII-1b exhibits several pairs of h-orthogonal vectors X, Y, each drawn as a pair of arrows issuing from the same point. All the vectors in this figure have zero x_1 coordinate, and so the x_1-axis is not shown. As you will be asked to verify in Exercise 3, in each case, the line bisecting the angle between X and Y is inclined 45° to the x_2-axis. (Of course, this would no longer be true in general when x_1 is non-zero.)

We shall make frequent use of the following identities.

$$<X,Y> = X_{ref} \cdot Y = X \cdot Y_{ref} \tag{32a}$$

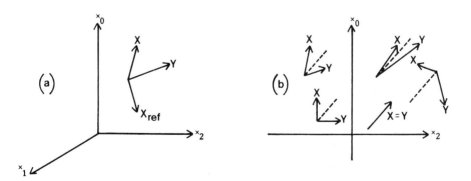

Figure VII - 1.

$$\mathbf{X} \cdot \mathbf{Y} = \langle \mathbf{X}_{ref}, \mathbf{Y} \rangle = \langle \mathbf{X}, \mathbf{Y}_{ref} \rangle \tag{32b}$$

$$\mathbf{X}_{ref} \times \mathbf{Y}_{ref} = - (\mathbf{X} \times \mathbf{Y})_{ref} \tag{32c}$$

$$\langle \mathbf{X} \times \mathbf{Y}, \mathbf{Z} \times \mathbf{W} \rangle = - \langle \mathbf{X}, \mathbf{Z} \rangle \langle \mathbf{Y}, \mathbf{W} \rangle + \langle \mathbf{Y}, \mathbf{Z} \rangle \langle \mathbf{X}, \mathbf{W} \rangle \tag{32d}$$

$\mathbf{X} \times \mathbf{Y}$ denotes the ordinary vector (or cross) product,

$$\mathbf{X} \times \mathbf{Y} = (x_1 y_2 - x_2 y_1, \, x_2 y_0 - x_0 y_2, \, x_0 y_1 - x_1 y_0)$$

Each of the above identities can be verified by computing both sides in terms of coordinates. However (32d) follows more quickly from rewriting the left-hand side as $-(\mathbf{X}_{ref} \times \mathbf{Y}_{ref}) \cdot (\mathbf{Z} \times \mathbf{W})$ and applying the identity

$$(\mathbf{A} \times \mathbf{B}) \cdot (\mathbf{C} \times \mathbf{D}) = (\mathbf{A} \cdot \mathbf{C})(\mathbf{B} \cdot \mathbf{D}) - (\mathbf{B} \cdot \mathbf{C})(\mathbf{A} \cdot \mathbf{D})$$

found in most books on vector analysis.

It is a basic property of the cross product that $\mathbf{X} \times \mathbf{Y}$ is e-orthogonal to both $\mathbf{X}$ and $\mathbf{Y}$. From (32 b,c) we then have the following important result.

Lemma VII-1

Let $\mathbf{X}$ and $\mathbf{Y}$ be any two linearly independent vectors of $\mathbb{R}^3$. Then $(\mathbf{X} \times \mathbf{Y})_{ref}$ is an h-normal for the plane spanned by $\mathbf{X}$ and $\mathbf{Y}$. If $\mathbf{A}$ is a vector h-orthogonal to both $\mathbf{X}$ and $\mathbf{Y}$, then $\mathbf{A}$ is a scalar multiple of $(\mathbf{X} \times \mathbf{Y})_{ref}$.

As we shall discover shortly, the Lobachevskian plane (with space constant k) can be represented on the surface

$$\langle \mathbf{X}, \mathbf{X} \rangle = - x_0^2 + x_1^2 + x_2^2 = - k^2$$

which is a hyperboloid of two sheets in $\mathbb{R}^3$ (Fig. VII-2). Rather than deal with antipodal point pairs, as we did in obtaining the projective plane from the sphere (cf. Section III-3), we shall simply throw away one of the sheets. The "lines" of our model will turn out to be plane sections of this hyperboloid made by planes through the origin. These are the analogues of great circles. In order for the congruence postulates to hold, we shall have to adopt a non-Euclidean definition of distance, based on the hyperbolic inner product (Section 3).

PRELIMINARIES

Before we get into details, we should point out that much of the sequel readily generalizes to higher dimensions. A point of $\mathbb{R}^{n+1}$ is a sequence $(x_0, x_1, x_2, \ldots, x_n)$, and the hyperbolic inner product is

$$\langle X, Y \rangle = -x_0 y_0 + x_1 y_1 + x_2 y_2 + \cdots + x_n y_n$$

The hypersurface $\langle X, X \rangle = -k^2$ (a "hyperhyperboloid")—with the appropriate definitions of distance and angle measure—is a model for n-dimensional Lobachevskian geometry. In the theory of relativity, this same inner product, with $n = 3$, plays a central role. There x_0 is a time coordinate, while x_1, x_2, and x_3 are space coordinates. For now however, $n = 2$.

Exercises VII-1

1. Prove: if P and Q are any two points of the plane with equation $A \cdot X = b$, then A and $P - Q$ are e-orthogonal. (Thus the line through P in the A direction is perpendicular to the line through P and Q. See Definition VI-1.)

2. Show that the perpendicular distance from the origin to the plane $A \cdot X = b$ is $|b|/(A \cdot A)^{1/2}$. (Hint: choose any point X of the plane and project the vector X onto the plane's normal line.)

3. Let X and Y be non-zero vectors in $\mathbb{R}^3$ which have zero x_1 coordinates, as in Figure VII-1b. Let these vectors be drawn emanating from the same point, e.g., the origin. Show that X and Y are h-orthogonal if and only if the line bisecting the angle formed by X and Y is inclined $\pm 45°$ to the x_2-axis. (More generally, if X and Y span a plane containing the x_3-axis, then X and Y are h-orthogonal if and only if the bisector of the angle they form is inclined $\pm 45°$ to the $x_1 x_2$-plane.)

4. Let X be a non-zero vector of $\mathbb{R}^3$. Geometrically describe the set of all vectors Y such that $\langle X, Y \rangle = 0$. Draw a sketch.

5. Verify the following identities:

   (i) $\langle X_{\text{ref}}, Y_{\text{ref}} \rangle = \langle X, Y \rangle$
   (ii) $\langle X \times Y, X \times Y \rangle = \langle X, Y \rangle^2 - \langle X, X \rangle \langle Y, Y \rangle$

6. (For the purpose of this exercise, which provides some motivation for the model to be constructed in subsequent sections, we assume Lobachevskian geometry is consistent.) Let r, θ be a system

of polar coordinates in the Lobachevskian plane with space constant k. (Polar coordinates are defined exactly as in Euclidean geometry.) To each point (r, θ) we shall associate the vector $\mathbf{X} = (x_0, x_1, x_2)$ of $\mathbb{R}^3$ given by

$$x_0 = k \cosh \frac{r}{k}$$

$$x_1 = k \sinh \frac{r}{k} \cos \theta$$

$$x_2 = k \sinh \frac{r}{k} \sin \theta$$

The numbers x_0, x_1, and x_2 are called the *Weierstrass coordinates* of the point with polar coordinates (r, θ).

(a) Show that $\mathbf{X}$ satisfies $\langle \mathbf{X}, \mathbf{X} \rangle = -k^2$, which is the equation of a hyperboloid of two sheets (Figure VII-2). Since $x_0 > 0$, $\mathbf{X}$ lies on the "upper sheet." Verify that Weierstrass coordinates furnish a one-to-one mapping of the Lobachevskian plane onto this upper sheet.

(b) Suppose $\mathbf{X} = (x_0, x_1, x_2)$ and $\mathbf{Y} = (y_0, y_1, y_2)$ are the Weierstrass coordinates of the points with polar coordinates (r, θ) and (r', θ'), respectively. Let d be the distance between the latter. Show that

$$\cosh \frac{d}{k} = -\frac{1}{k^2} \langle \mathbf{X}, \mathbf{Y} \rangle$$

[Hint: use equation (25) of Section VI-5.]

(c) For a given line in the Lobachevskian plane, let p be the perpendicular distance from the origin, and suppose that α is the angle between the ray $\theta = 0$ and the line's normal directed from the origin to the line (see figure).

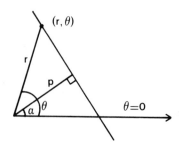

PRELIMINARIES

The numbers p and α are called polar coordinates of the line. From (21) of Section VI-5, a point (r, θ) is on the line (p, α) if and only if

$$\tanh \frac{p}{k} = \tanh \frac{r}{k} \cos(\theta - \alpha)$$

Show that this equation is equivalent to $\langle X, \ell \rangle = 0$, where $\ell = (\lambda_0, \lambda_1, \lambda_2)$ is given by

$$\lambda_0 = k \sinh \frac{p}{k}$$

$$\lambda_1 = k \cosh \frac{p}{k} \cos \alpha$$

$$\lambda_2 = k \cosh \frac{p}{k} \sin \alpha$$

These are the *Weierstrass coordinates* of the line (p, α). Verify that $\langle \ell, \ell \rangle = k^2$. Consequently, since $\langle X, \ell \rangle = 0$ is a linear equation in x_0, x_1, and x_2, lines of the Lobachevskian plane correspond to sections of the hyperboloid made by planes through the origin.

7. Sketch and give h-normal vectors for each of the following planes. Include the h-normals on your sketches.

   (i) $x_0 = x_2/2$
   (ii) $x_0 + x_2 = 0$
   (iii) $2x_0 + 3x_1 = 6$

2. H^2

We are now prepared to construct our model, and thereby prove that Lobachevskian geometry is consistent. As you know, we must specify interpretations for the primitive terms (point, line, lies on, between, and congruent) and verify Hilbert's plane axioms from groups I, II, III, and V, and the Lobachevskian parallel postulate.

A "point" is defined to be a point (or vector) X of $\mathbb{R}^3$ such that $\langle X, X \rangle = -k^2$ and $x_0 > 0$. (The constant k will turn out to be the space constant.) Such points comprise the "upper sheet" of a 2-sheeted hyperboloid. We shall call this sheet H^2. The asymptotic cone of H^2 consists of all X of $\mathbb{R}^3$ satisfying $\langle X, X \rangle = 0$ (Fig. VII-2). This is sometimes called the *light cone*, a term which is borrowed from relativity theory, where a similar cone is used to describe

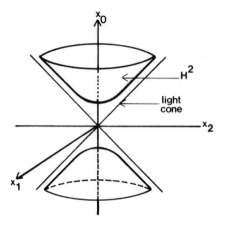

Figure VII - 2.

graphically how light travels. You can easily verify that points **X** in the interior of the light cone (e.g., points of H^2) satisfy $\langle X,X\rangle < 0$, while points **Y** outside the light cone satisfy $\langle Y,Y\rangle > 0$.

A "line" is defined to be the intersection of H^2 with a plane through the origin of $\mathbb{R}^3$. It is known from Euclidean solid geometry that such a plane section is one branch of a hyperbola (Fig. VII-3). (The other branch lies on the discarded lower sheet of $\langle X,X\rangle = -k^2$.) Now a plane through the origin is given by an equation of the form $\langle X,\ell\rangle = 0$, where ℓ is an h-normal of the plane. In the present case, the point ℓ lies outside the light cone (why?), and so $\langle \ell,\ell\rangle > 0$. Since any scalar multiple of ℓ determines the same plane $\langle X,\ell\rangle = 0$,

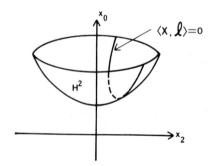

Figure VII - 3.

we shall always assume that ℓ has been chosen so that $\langle\ell,\ell\rangle = k^2$, which means that ℓ is a point of the single-sheeted hyperboloid with equation $\langle X,X\rangle = k^2$, depicted in Figure VII-4. Because ℓ and $-\ell$ determine the same plane $\langle X,\ell\rangle = 0$, we must identify (consider as equivalent) each point ℓ of the single-sheeted hyperboloid and its opposite or antipodal point, $-\ell$.

In summary, a "line" is a section of H^2 by a plane through the origin of $\mathbb{R}^3$. There is a vector ℓ, unique up to sign, such that (a) $\langle\ell,\ell\rangle = k^2$, and (b) the line consists of all X of H^2 such that $\langle X,\ell\rangle = 0$. "Lines" are thus in one-to-one correspondence with antipodal point pairs $(\ell,-\ell)$ on the hyperboloid $\langle X,X\rangle = k^2$.

From now on, we shall refer to a "line" either by giving its equation, $\langle X,\ell\rangle = 0$, or simply by naming its vector ℓ (as in "the line ℓ"). In order to distinguish between (Lobachevskian) "lines," which are plane sections of H^2, and ordinary Euclidean lines in $\mathbb{R}^3$, we shall henceforth refer to the former as *lines* (without quotation marks) and to the latter as *straight lines*.

A point X of H^2 *lies on* the line ℓ if and only if $\langle X,\ell\rangle = 0$. Two points X and Y of H^2 lie on a unique line since X, Y, and the origin determine a unique plane through the latter. Thus, Hilbert's postulates I, 1, 2, and 3 clearly hold.

We saw earlier that a line in this model is one branch of a hyperbola. Given three distinct points A, B, and C on the line, we will say C is *between* A and B if, when the branch is traversed in either

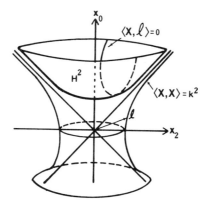

Figure VII - 4.

direction, the points are encountered in the order ACB or BCA. This can be expressed more succinctly in vector terminology. Since A, B, and C are coplanar, any one of them is a linear combination of the other two. C is between A and B if and only if C is a linear combination of A and B with positive coefficients:

$$C = \lambda A + \mu B, \quad \lambda, \mu > 0$$

Axioms II, 1, 2, and 3 of Hilbert are now obvious, and II,4 follows from its equivalent, the Separation Axiom, which holds here because a plane $\langle X, \ell \rangle = 0$, where $\langle \ell, \ell \rangle = k^2$, divides H^2 into two sets satisfying the requisite properties (check this). Segments, rays, angles, and triangles are defined in terms of betweenness, as in Appendix B. Dedekind's Postulate, which is equivalent to Hilbert's postulate group V, will be proved in Exercise 5, below.

Exercises VII-2

1. Let P and Q be points of H^2 and assume ℓ and m are lines of H^2. Prove: (a) if ℓ and m intersect in P, then P is a scalar multiple of $(\ell \times m)_{\text{ref}}$; (b) if ℓ joins P to Q, then ℓ is a scalar multiple of $(P \times Q)_{\text{ref}}$.

2. *Concurrency and collinearity*: (a) Prove that the lines ℓ, m, and n are concurrent if and only if $\ell \cdot m \times n = 0$. (b) Prove that the points X, Y, and Z of H^2 are collinear if and only if $X \cdot Y \times Z = 0$. (Note: because of the identity

$$A \cdot B \times C = \det \begin{pmatrix} a_0 & a_1 & a_2 \\ b_0 & b_1 & b_2 \\ c_0 & c_1 & c_2 \end{pmatrix}$$

each of the above conditions can be expressed as the vanishing of a certain determinant.)

3. Let $P = (p_0, p_1, \ldots, p_n)$ and $Q = (q_0, q_1, \ldots, q_n)$ be points of $\mathbb{R}^{n+1}$ satisfying

(i) $p_0, q_0 > 0$, (ii) $\langle P, P \rangle < 0$, (iii) $\langle Q, Q \rangle \leq 0$

Prove that $\langle P, Q \rangle < 0$.

[Hint: Let **X** and **Y** be the n-dimensional vectors formed by the last n coordinates of **P** and **Q**:

$$X = (p_1, p_2, \ldots, p_n), \quad Y = (q_1, q_2, \ldots, q_n)$$

Suppose, contrary to what we must prove, that $\langle P,Q\rangle = - p_0 q_0 + X \cdot Y \geq 0$. Then, by (i), we have $0 < p_0 q_0 \leq X \cdot Y$. Use the Cauchy-Schwartz inequality (cf. Section 1) to deduce a contradiction from (ii) and (iii).]

4. Let **P** and **Q** be points of H^2. Show that $\langle P,Q\rangle \leq -k^2$, and that equality holds if and only if $P = Q$.
[Hint:

$$\frac{\langle P,Q\rangle^2}{k^4} - 1 = \frac{\langle P,Q\rangle^2}{\langle P,P\rangle\langle Q,Q\rangle} - 1 = \frac{\langle P,Q\rangle^2 - \langle P,P\rangle\langle Q,Q\rangle}{k^4}$$

$$= \frac{\langle P \times Q, P \times Q\rangle}{k^4}$$

Now use Exercises 1(b) and 3.]

5. Let **A** and **B** be the intersections of the asymptotes of the hyperbola $y^2 - x^2 = 1$ (in the Cartesian plane) with the hyperbola's tangent line at $(0,1)$. (a) Show there is a one-to-one order-preserving correspondence between the points of segment $\overline{AB}$ and the points of the "upper" branch of the hyperbola. (b) Explain briefly how it follows from the above that Dedekind's Postulate holds in H^2.

3. DISTANCE IN H^2

According to Exercise 4 of the preceding section, if **P** and **Q** are points of H^2, then $\langle P,Q\rangle \leq -k^2$, and equality holds if and only $P = Q$. Consequently, the right hand side of (33) below is at least unity and therefore in the range of the hyperbolic cosine.

Definition VII-2

The (Lobachevskian) *distance* between two points **P** and **Q** of H^2 is defined to be the unique non-negative number $d = d_{PQ}$ satisfying

$$\cosh \frac{d}{k} = -\frac{1}{k^2} \langle P,Q\rangle \qquad (33)$$

We define congruence of segments in terms of distance. If **A**, **B**, **C**, and **D** are points of H^2, we shall say $\overline{AB}$ is congruent to $\overline{CD}$ (written $AB = CD$) if $d_{AB} = d_{CD}$.

Let us now verify axiom III,1. Given **A**, **B**, and **P** on H^2, with **P** on line ℓ, we must show that on each side of **P** there is a point **Q** on ℓ such that $PQ = AB$. Let d be the distance from **A** to **B**, i.e.,

$$\cosh \frac{d}{k} = -\frac{1}{k^2} <A,B>$$

The points **X** whose distance from **P** equals d satisfy

$$\cosh \frac{d}{k} = -\frac{1}{k^2} <P,X> \tag{34}$$

This is the equation of a plane with e-normal P_{ref}, which lies inside the light cone (Fig. VII-5). It is known from solid geometry that such a plane intersects H^2 in an ellipse—unless it is tangent to H^2 at one point or misses it altogether. We will next show the intersection is indeed an ellipse by proving that **P** lies below the plane, i.e., on the origin side of it.

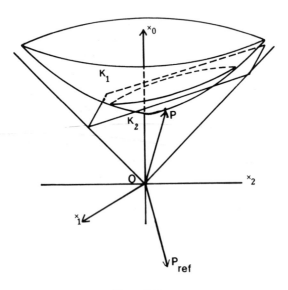

Figure VII - 5.

DISTANCE IN H²

The plane (34) determines two half-spaces, namely

$$K_1 = \left\{ X \in \mathbb{R}^3 \mid \cosh \frac{d}{k} < -\frac{1}{k^2} <P,X> \right\}$$

$$K_2 = \left\{ X \in \mathbb{R}^3 \mid \cosh \frac{d}{k} > -\frac{1}{k^2} <P,X> \right\}$$

K_1 is the half-space above the plane, since points of the form $X = (x_0, 0, 0)$ satisfy

$$-\frac{1}{k^2} <P,X> = -\frac{1}{k^2}(-p_0 x_0) = \frac{p_0}{k^2} x_0 > \cosh \frac{d}{k}$$

if x_0 is sufficiently large. Since

$$\cosh \frac{d}{k} > 1 = -\frac{1}{k^2}(-k^2) = -\frac{1}{k^2} <P,P>$$

P is in K_2 and so lies below the plane, as claimed. Thus the plane (34) intersects H² in an ellipse. Therefore any line through **P** meets this ellipse in exactly two points, one on each side of **P**. This establishes Postulate III,1. As a by-product, we have a characterization of circles in our model.

Theorem VII-3
   In the model H², a Lobachevskian circle with center **P** and radius r consists of all **X** of H² satisfying

$$\cosh \frac{r}{k} = -\frac{1}{k^2} <X,P> = \frac{1}{k^2}(x_0 p_0 - x_1 p_1 - x_2 p_2)$$

This is a plane section whose h-normal is the vector from the origin to **P**.

   Figure VII-6 illustrates several concentric circles with center **P**. They are ellipses lying in parallel planes.

   Postulate III,2 is obvious because equality of distance is transitive. Before establishing III,3 we need a preliminary result that will enable us to deduce the additivity of distance (Theorem VII-5, below).

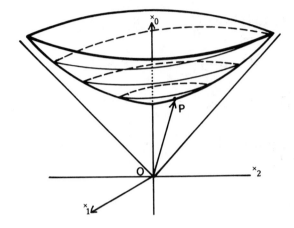

Figure VII - 6.

Lemma VII-4

Let **A**, **B**, and **C** be collinear points of H^2 such that **C** is between **A** and **B**. Let λ and μ be the unique positive numbers such that

$$C = \lambda A + \mu B \tag{35}$$

Then

$$<A,C>^2 - k^4 = \mu^2 (<A,B>^2 - k^4) \tag{36a}$$

$$<B,C>^2 - k^4 = \lambda^2 (<A,B>^2 - k^4) \tag{36b}$$

Proof. From (35) we have

$$<C,C> = -\lambda^2 k^2 + 2\lambda\mu<A,B> - \mu^2 k^2 \tag{37a}$$

$$<A,C> = -\lambda k^2 + \mu<A,B> \tag{37b}$$

$$<B,C> = \lambda<A,B> - \mu k^2 \tag{37c}$$

Therefore,

DISTANCE IN H²

$$\langle A,C\rangle^2 - k^4 = \langle A,C\rangle^2 + k^2\langle C,C\rangle$$
$$= \lambda^2 k^4 - 2\lambda\mu k^2\langle A,B\rangle + \mu^2\langle A,B\rangle^2$$
$$+ k^2(-\lambda^2 k^2 + 2\lambda\mu\langle A,B\rangle - \mu^2 k^2)$$
$$= \mu^2(\langle A,B\rangle^2 - k^4)$$

(36b) is proved similarly.

Theorem VII-5

Let **A**, **B**, and **C** be collinear points of H². Then if **C** is between **A** and **B**,

$$d_{AB} = d_{AC} + d_{CB}.$$

Proof. We can write $C = \lambda A + \mu B$, where $\lambda, \mu > 0$.

$$\cosh\frac{d_{AC}+d_{CB}}{k} = \cosh\frac{d_{AC}}{k}\cosh\frac{d_{CB}}{k} + \sinh\frac{d_{AC}}{k}\sinh\frac{d_{CB}}{k}$$
$$= \cosh\frac{d_{AC}}{k}\cosh\frac{d_{CB}}{k} + (\cosh^2\frac{d_{AC}}{k} - 1)^{1/2}$$
$$\times (\cosh^2\frac{d_{CB}}{k} - 1)^{1/2}$$
$$= \frac{1}{k^4}\langle A,C\rangle\langle C,B\rangle + \frac{1}{k^4}(\langle A,C\rangle^2 - k^4)^{1/2}$$
$$\times (\langle B,C\rangle^2 - k^4)^{1/2}$$

We now use (36) and (37) to rewrite the last expression in terms of **A**, **B**, λ, and μ. This produces

$$k^4\cosh\frac{d_{AC}+d_{CB}}{k} = (-\lambda k^2 + \mu\langle A,B\rangle)(\lambda\langle A,B\rangle - \mu k^2)$$
$$+ \lambda\mu(\langle A,B\rangle^2 - k^4)*$$
$$= \langle A,B\rangle(-\lambda^2 k^2 + 2\lambda\mu\langle A,B\rangle - \mu^2 k^2)$$
$$= \langle A,B\rangle\langle C,C\rangle = -k^2\langle A,B\rangle = k^4\cosh\frac{d_{AB}}{k}.$$

---
*The positivity of λ, μ appears here in $(\lambda^2\mu^2)^{1/2} = \lambda\mu$.

Therefore $d_{AC} + d_{CB} = d_{AB}$, and the theorem is proved.

It is now easy to verify Postulate III-3. If C is between A and B, and C' is between A' and B', and if AC = A'C' and CB = C'B', we have

$$d_{AB} = d_{AC} + d_{CB} = d_{A'C'} + d_{C'B'} = d_{A'B'},$$

and so AB = A'B'.

4. PARAMETRIC EQUATION OF A LINE

Sometimes it will be useful to think of a Lobachevskian line ℓ as the path traversed by a moving point $X(t) = (x_0(t), x_1(t), x_2(t))$ tracing out a plane section of H^2. (We assume differentiability of the coordinate functions. The parameter t may be thought of as time.) The derivative,

$$X'(t) = (dx_0/dt, dx_1/dt, dx_2/dt)$$

is called the *velocity vector* and is tangent to ℓ and therefore to the surface H^2. Let $s = s(t)$ denote the (Lobachevskian) distance from $X(0)$ to $X(t)$. (Choose a direction on the line so that distance is reckoned positively in this direction and negatively in the opposite direction.)

Lemma VII-6

$$(ds/dt)^2 = \langle X'(t), X'(t) \rangle.$$

Proof. Note that as long as $X'(t) \neq 0$, $\langle X'(t), X'(t) \rangle$ will be positive, since any vector tangent to H^2, when parallel-translated to the origin, points outside the light cone. We shall establish the lemma at an arbitrary point $X(t_0)$ on the line. Let $\Delta s = s(t) - s(t_0)$. [By Theorem VII-5, this is the distance from $X(t_0)$ to $X(t)$.] Then $d/dt(\Delta s) = s'(t)$. Repeated differentiation of

$$\cosh \frac{\Delta s}{k} = -\frac{1}{k^2} \langle X(t), X(t_0) \rangle$$

PARAMETRIC EQUATION OF A LINE

with respect to t gives

$$\frac{s'(t)}{k} \sinh \frac{\Delta s}{k} = -\frac{1}{k^2} \langle X'(t), X(t_0) \rangle$$

$$\frac{s''(t)}{k} \sinh \frac{\Delta s}{k} + \frac{(s'(t))^2}{k^2} \cosh \frac{\Delta s}{k} = -\frac{1}{k^2} \langle X''(t), X(t_0) \rangle$$

In the last equation, set $t = t_0$, so that $\Delta s = 0$. Then multiply by k^2. The result is

$$\left(\frac{ds}{dt}\right)^2_{t=t_0} = -\langle X''(t_0), X(t_0) \rangle = \langle X'(t_0), X'(t_0) \rangle$$

[The last equality results from twice differentiating $\langle X(t), X(t) \rangle = -k^2$.] Since t_0 was arbitrary, the lemma is proved.

If, as we shall henceforth assume, distance itself is chosen as the parameter ($t = s$), then the line is said to be *parametrized by distance*. Since in this case $ds/dt = 1$, $X'(s)$ satisfies $\langle X'(s), X'(s) \rangle = 1$. Now since our line is a plane section of H^2, then for any s, the vector $X(s)$ lies in the plane (through the origin) spanned by the vectors $P = X(0)$ and $T = kX'(0)$. [It will be more convenient to work with T rather than $X'(0)$. Picture T as translated to the origin as in Fig. VII-7.]

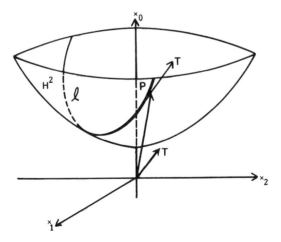

Figure VII - 7.

$X(s)$ therefore may be written in the form $X(s) = \alpha(s)P + \beta(s)T$, where α and β are differentiable real-valued functions. P and T satisfy

$$\langle P,P \rangle = -k^2, \quad \langle T,T \rangle = k^2, \quad \langle P,T \rangle = 0 \tag{38}$$

(the last because $\langle X, X' \rangle = 0$ for all s). From these and the identities $\langle X,X \rangle = -k^2$, $\langle X',X' \rangle = 1$, we can deduce that $\alpha^2 - \beta^2 = 1$, $\beta'^2 - \alpha'^2 = 1/k^2$, where $\alpha(0) = 1$, $\beta(0) = 0$.

The functions $\alpha(s) = \cosh s/k$ and $\beta(s) = \sinh s/k$ satisfy these equations, and it can be shown that these are the only solutions (Exercise 2). We have therefore arrived at the parametric equation

$$\boxed{X(s) = (\cosh s/k)P + (\sinh s/k)T} \tag{39}$$

for the Lobachevskian line satisfying $X(0) = P$, $X'(0) = (1/k)T$. Conversely, if (38) holds, (39) represents a line through P with tangent vector T (check this).

Exercises VII-4

1. Let $X(s) = (\cosh s/k)P + (\sinh s/k)T$ be the distance parametrization of the Lobachevskian line ℓ (where, as always, $\langle \ell, \ell \rangle = k^2$). Show that

$$\ell = \pm \frac{1}{k}(P \times T)_{\text{ref}}$$

2. Show that $\alpha(s) = \cosh s/k$, $\beta(s) = \sinh s/k$ are the only solutions to the system of equations

$$\alpha^2 - \beta^2 = 1, \quad (\beta')^2 - (\alpha')^2 = 1/k^2$$

subject to the initial conditions

$$\alpha(0) = 1, \quad \beta(0) = 0, \quad \beta'(0) > 0$$

(Hint: From the first equation, $\alpha \alpha' = \beta \beta'$. Multiply the second equation by α^2 and deduce that $(\beta')^2 = \alpha^2/k^2$, or $\beta' = +\alpha/k$, the choice of sign being dictated by the initial conditions. Then show $\alpha' = \beta/k$. You

PARAMETRIC EQUATION OF A LINE

have thus shown that solutions to the original system are also solutions to the linear system $\alpha' = \beta/k$, $\beta' = \alpha/k$. The latter has unique solutions because of the following theorem from differential equation theory.

Theorem. Let

$y_i' = a_{i1}(x)y_1 + a_{i2}(x)y_2 + \cdots + a_{in}(x)y_n$, $i = 1,2,\ldots,n$, be a linear system of differential equations, where the functions $a_{ij}(x)$ are continuous real-valued functions defined on a certain open interval. Given any real numbers $b_1, b_2, \ldots, b_n$ and any point x_0 in the given interval, there exist unique functions $y_1, y_2, \ldots, y_n$ which are solutions to the above system on the entire interval and which satisfy the initial conditions $y_i(x_0) = b_i$ for each i.

(See Pontryagin [37, p. 22], Cole [9, p. 42], or White [51, pp. 227, 233].)

5. ANGLES

Let $\triangle ABC$ be an arbitrary triangle on H^2. Label the sides and angles as in Figure VII-8. Parametrize $\overleftrightarrow{CA}$ and $\overleftrightarrow{CB}$ by distance, so that we may write

$$A = (\cosh \frac{b}{k})C + (\sinh \frac{b}{k})T$$

$$B = (\cosh \frac{a}{k})C + (\sinh \frac{a}{k})U$$
(40)

where $<T,T> = <U,U> = k^2$, $<C,T> = <C,U> = 0$, and T and U are respectively tangent at C to the rays $\overrightarrow{CA}$ and $\overrightarrow{CB}$.

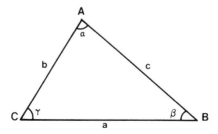

Figure VII - 8.

Now, from (40),

$$\cosh\frac{c}{k} = -\frac{1}{k^2}\langle A,B\rangle = \cosh\frac{a}{k}\cosh\frac{b}{k} - \sinh\frac{a}{k}\sinh\frac{b}{k}\frac{\langle T,U\rangle}{k^2}$$

If we were to define angle measure so that

$$\cos\gamma = \frac{\langle T,U\rangle}{k^2} = \frac{\langle T,U\rangle}{(\langle T,T\rangle\langle U,U\rangle)^{1/2}} \qquad (41)$$

then the Lobachevskian Law of Cosines, (25) of Section VI-5, would hold in our model. This is exactly what we shall do, but we must first show that $\langle T,U\rangle/k^2$ is in the range of the cosine function, i.e., between -1 and 1. To see this, note first that since T and U are tangent to the surface H^2 at C, the vector $T \times U$ is an e-normal to this surface at C and so, if translated to the origin, points into the interior of the light cone. Accordingly, $\langle T\times U, T\times U\rangle < 0$. Therefore,

$$\frac{\langle T,U\rangle^2}{\langle T,T\rangle\langle U,U\rangle} - 1 = \frac{\langle T,U\rangle^2 - \langle T,T\rangle\langle U,U\rangle}{k^4} = \frac{\langle T\times U, T\times U\rangle}{k^4}$$
$$< 0,$$

and so the right side of (41) is indeed between -1 and 1.

Definition VII-7

The angle formed at a point P of H^2 by two rays

$$X(s) = (\cosh\frac{s}{k})P + (\sinh\frac{s}{k})T, \quad s \geq 0$$

$$Y(s) = (\cosh\frac{t}{k})P + (\sinh\frac{t}{k})U, \quad t \geq 0$$

where $\langle P,T\rangle = \langle P,U\rangle = 0$, $\langle T,T\rangle = \langle U,U\rangle = k^2$, has measure θ given by

$$\cos\theta = \frac{\langle T,U\rangle}{(\langle T,T\rangle\langle U,U\rangle)^{1/2}} \qquad (42)$$

An alternative formula is derived in Exercise 1. Since the Lobachevskian Law of Cosines now holds, it is a simple matter to verify Hilbert's Postulate III,5. Suppose $\triangle A'B'C'$ is a second triangle (labeled as in Fig. VII-8, but with primes) and that $a = a'$, $b = b'$, and $\gamma \cong \gamma'$. We must prove $\alpha \cong \alpha'$. From the assumed congruences and (25) of Section VI-5, we obtain

$$\cosh\frac{c'}{k} = \cosh\frac{a'}{k}\cosh\frac{b'}{k} - \sinh\frac{a'}{k}\sinh\frac{b'}{k}\cos\gamma'$$

$$= \cosh\frac{a}{k}\cosh\frac{b}{k} - \sinh\frac{a}{k}\sinh\frac{b}{k}\cos\gamma = \cosh\frac{c}{k}$$

and so $c = c'$. But then, by a permuted version of VI-(25), we have

$$\cos\alpha' = \frac{\cosh\frac{b'}{k}\cosh\frac{c'}{k} - \cosh\frac{a'}{k}}{\sinh\frac{b'}{k}\sinh\frac{c'}{k}}$$

$$= \frac{\cosh\frac{b}{k}\cosh\frac{c}{k} - \cosh\frac{a}{k}}{\sinh\frac{b}{k}\sinh\frac{c}{k}} = \cos\alpha$$

Therefore $\alpha \cong \alpha'$ and III,5 is established.

To verify III,4, suppose we are given a ray $\overrightarrow{AB}$ in H^2 and a number θ such that $0 < \theta < 2\pi$. (θ is the radian measure of the angle to be copied.) Let

$$\alpha(s) = (\cosh\frac{s}{k})A + (\sinh\frac{s}{k})T, \quad s \geq 0$$

be the distance parametrization of $\overrightarrow{AB}$, where $\langle A,T\rangle = 0, \langle T,T\rangle = k^2$ (s increases in the direction from A to B). Any ray $\overrightarrow{AC}$ has a similar parametrization,

$$\beta(t) = (\cosh\frac{t}{k})A + (\sinh\frac{t}{k})U, \quad t \geq 0 \tag{43}$$

where $\langle A,U\rangle = 0, \langle U,U\rangle = k^2$. We must show that in a given half-plane of $\overleftrightarrow{AB}$ there exists a unique ray $\overrightarrow{AC}$ for which $<T,U>/k^2 = \cos\theta$.

First we construct the perpendicular to $\overleftrightarrow{AB}$ at A. Let $<X,\ell> = 0$ be the equation of $\overleftrightarrow{AB}$, where $<\ell,\ell> = k^2$. Then since

$$<A,\ell> = <T,\ell> = 0, \quad <\ell,\ell> = k^2$$

the curve

$$\gamma(u) = (\cosh \tfrac{u}{k})A + (\sinh \tfrac{u}{k})\ell, \quad u \geq 0$$

is a Lobachevskian ray with tangent vector ℓ at A, by the results of Section 4. Moreover, since $<T,\ell> = 0$, this ray is perpendicular to $\overleftrightarrow{AC}$ at A (Definition VII-7). By multiplying ℓ by -1 if necessary, we may assume the ray is directed into the desired half-plane of $\overleftrightarrow{AC}$.

Finally, if we define U by

$$U = (\cos\theta)T + (\sin\theta)\ell$$

then, as you can check, $<A,U> = 0$, $<U,U> = k^2$, and $<T,U>/k^2 = \cos\theta$. Accordingly, the ray (43) forms with $\overrightarrow{AB}$ an angle of measure θ. This establishes III,4.

Exercises VII-5

1. Let $<X,\ell> = 0$ and $<X,m> = 0$ be the respective equations of the lines $X(s)$ and $Y(s)$ of Definition VII-7. Since $<T,P> = <T,\ell> = 0$ and $<U,P> = <U,m> = 0$, T and U are scalar multiples of $(P \times \ell)_{ref}$ and $(P \times m)_{ref}$, respectively. Prove that

$$\cos^2\theta = \frac{<\ell,m>^2}{<\ell,\ell><m,m>}$$

(Consequently, lines ℓ and m are perpendicular if and only if $<\ell,m> = 0$.)

2. (Cf. Exercise 1 of Section VII-2.) Let P and Q be points of H^2, and assume ℓ and m are lines of H^2. Prove: (a) if ℓ and m intersect in P, forming vertical angles of measure θ and $\pi - \theta$, then

ANGLES

$$P = \pm \frac{1}{k \sin \theta} (\ell \times m)_{ref}$$

(b) if ℓ joins P to Q, then

$$\ell = \pm \frac{1}{k \sinh(d/k)} (P \times Q)_{ref}$$

where d is the distance from P to Q.
(Hint: if $<A,A> = <B,B> = \pm k^2$, then $(A \times B, A \times B) = <A,B>^2 - <A,A><B,B> = k^4 [<A,B>^2/<A,A><B,B> - 1]$.)

6. THE HOMOGENEOUS REPRESENTATION

Each point X of H^2 determines a straight line $\overleftrightarrow{OX}$ passing through the origin and the interior of the light cone. Conversely, a straight line passing through the light cone's interior meets H^2 in a unique point. Therefore, we may represent the Lobachevskian plane either as the set of all points of H^2 or as the set of all lines through the origin which pass inside the light cone. Since these two sets are naturally in one-to-one correspondence, these two viewpoints are entirely equivalent (the two interpretations are isomorphic). In the sequel, for any point $X \neq O$, we shall write $[X]$ for the straight line $\overleftrightarrow{OX}$, i.e., for the collection of all real multiples of the vector X. Any point Y ($\neq O$) on such a line—being a non-zero multiple of a point of H^2—satisfies $<Y,Y> < 0$.

Equation (33), giving the distance between two points P and Q on H^2, may be rewritten

$$\cosh \frac{d}{k} = \frac{-<P,Q>}{(<P,P><Q,Q>)^{1/2}} \qquad (44)$$

since $<P,P> = <Q,Q> = -k^2$. The right hand side of this equation is homogeneous, that is to say, its value is unchanged if we multiply P by a non-zero constant, or if we multiply Q by a non-zero constant, or both. Therefore, we may take (44) as the equation for the distance between two straight lines $[P]$ and $[Q]$, where P and Q are points within the light cone. By homogeneity, the value of d is determined solely by the two lines and not by the particular choices of P and Q on these lines.

In summary then, in the homogeneous interpretation a Lobachevskian point is a straight line, [P], where $<P,P> < 0$. The components of P are called *homogeneous coordinates* of [P]. A Lobachevskian line is a plane passing through the origin and the interior of the light cone.

We will show next that the Lobachevskian parallel postulate holds in the Weierstrass Model by verifying this postulate for the homogeneous interpretation. Here a Lobachevskian line and a point not on it are represented by a plane α which passes through the origin and the light cone's interior, and by a straight line [P] not lying on α. Although every plane containing [P] must intersect α in a straight line through the origin, this straight line need not meet the interior of the light cone. In fact, there are infinitely many planes which contain [P] but whose intersections with α fail to meet H^2. Each such plane represents a line through the given point which does not meet the given line.

We have now proved that the Weierstrass interpretation is a model for Lobachevskian geometry, and therefore that this geometry is consistent. In the remaining sections, we shall look at other features of this model and relate it to the Klein-Beltrami Disk Model (Exercise 2 of Section 9).

7. PARALLELS AND HOROCYCLES

Suppose $X(s) = (\cosh s/k)P + (\sinh s/k)T$ is the equation of a line in H^2 parametrized by distance [equations (38) are assumed]. In the homogeneous representation, we would write [X(s)], and of course if f(s) is any never-zero function on $\mathbb{R}$, then [f(s)X(s)] is the same line. By letting $s \to +\infty$ in $2e^{-s/k}X(s)$ and $s \to -\infty$ in $2e^{+s/k}X(s)$, we obtain

$$\lim_{s \to +\infty} [X(s)] = [P + T], \quad \lim_{s \to -\infty} [X(s)] = [P - T]$$

(To verify these, express cosh and sinh in terms of exponentials. By $[X(s)] \to [P + T]$ we mean that the ordinary Euclidean angle between these lines approaches zero as $s \to +\infty$, and similarly for the other limit.) [P + T] and [P - T] are therefore to be viewed as the two ideal points on the line [X(s)].

PARALLELS AND HOROCYCLES

Since $\langle P + T, P + T\rangle = \langle P - T, P - T\rangle = 0$, we see that ideal points are represented in our model by the elements of the light cone. In fact $[P + T]$ and $[P - T]$ are the asymptotes of the hyperbola of which one branch is the Lobachevskian line $X(s)$. Accordingly, if $A\ (\neq O)$ is a point of the light cone, two parametrized lines $X(s)$ and $Y(s)$ are parallel in the direction $[A]$ if

$$\lim_{s \to +\infty} [X(s)] = \lim_{s \to +\infty} [Y(s)] = [A]$$

or, equivalently, if $[A]$ is the intersection of the planes containing these lines. Moreover, if P is any point of H^2, then the plane spanned by P and A—denote this plane $[P,A]$—intersects H^2 in the unique Lobachevskian line through P and the ideal point $[A]$. If P is allowed to vary over all of H^2, then the plane section $[P,A] \cap H^2$ varies over the pencil of lines parallel in the direction $[A]$.

Let us now find the equation of a horocycle. If $X(s) = [\cosh(s/k)]\,P + [\sinh(s/k)]\,T$ is a parametrized line, the circle through P with center $X(s)$ is the collection of all points $X \in H^2$ satisfying $\langle X(s), X\rangle = \langle X(s), P\rangle$, or $\langle X - P, X(s)\rangle = 0$. As explained in Exercise 7 of Section V-12, as the radius s becomes infinite, the circle approaches a horocycle in the limit. Using this, we can now derive the equation of a horocycle.

Rewriting the circle's equation in the equivalent form $\langle X - P, 2e^{-s/k}X(s)\rangle = 0$ and letting $s \to +\infty$ (so that $2e^{-s/k}X(s) \to P + T$), we have, in the limit

$$\langle X - P, A\rangle = 0 \tag{45}$$

where A is any convenient point (other than the origin) on $[P + T]$. (45) is the equation of the horocycle through P with center $[A]$ on the absolute (see Fig. VII-9). Thus, horocycles too are plane sections of H^2, and it can be shown that such sections are parabolas.

From now on, assume A has been chosen so as to lie on the top nappe of the light cone, i.e. $a_0 > 0$. By Exercise 3, Section VII-2, $\langle P, A\rangle < 0$, and we will "standardize" by replacing A in (45) by the vector $-A/\langle P, A\rangle$ (also on the top nappe). Assuming the name A now denotes the latter vector, (45) takes the "standard" form

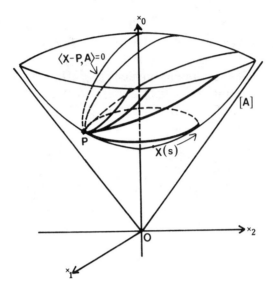

Figure VII - 9.

$$\langle A, X \rangle = -a_0 x_0 + a_1 x_1 + a_2 x_2 = -1 \tag{46}$$

Theorem VII-8

A horocycle in the Lobachevskian plane is represented on H^2 by a plane section whose h-normal is an element of the light cone. Specifically, the horocycle through $P \in H^2$ with center $[A]$ on the absolute consists of all $X \in H^2$ satisfying $\langle X, A \rangle = \langle P, A \rangle$. There is a unique choice of A on the upper nappe of the light cone for which the equation takes the form $\langle A, X \rangle = -1$.

Proof. We have proved all but the uniqueness of A in the standard form. If $\langle A, X \rangle = -1$ and $\langle B, X \rangle = -1$ are equations for the same horocycle, then for any point X on this horocycle, $\langle A - B, X \rangle = 0$. If $A \neq B$, the latter is the equation of a plane through the origin, and so determines a Lobachevskian line that contains the given horocycle—an impossibility (Theorem V-27). Hence $A = B$.

In Exercise 3, you will be asked to show that the set of all $X \in H^2$ such that $\langle A, X \rangle = -1$ intersects all the lines through the ideal point $[A]$ orthogonally. This should convince you that the limit of circles is indeed a horocycle.

PARALLELS AND HOROCYCLES

Note that two horocycles, $\langle A,X \rangle = -1$ and $\langle B,X \rangle = -1$ (A, B on the top nappe of the light cone) are *concentric* if and only if $B = \lambda A$ for some $\lambda > 0$. In Exercise 1, you will show that $k|\ln \lambda|$ is the distance between these horocycles.

Exercises VII-7

1. *Distance between concentric horocycles.* Let $\langle A,X \rangle = -1$ and $\langle B,X \rangle = -1$ be the equations of two concentric horocycles on H^2, where $B = \lambda A$, $1 \neq \lambda > 0$. Suppose a common axis intersects the horocycles in P and Q, so that $\langle A,P \rangle = \langle B,Q \rangle = -1$. Since this axis is the section of H^2 by the plane [P,A], we may express Q as $Q = \alpha P + \beta A$, where $\alpha, \beta \in \mathbb{R}$. The distance between the horocycles, d, is given by

$$\cosh \frac{d}{k} = -\frac{1}{k^2} \langle P,Q \rangle$$

We shall treat d as a signed distance, positive if the direction from P to Q coincides with the direction [A], negative otherwise. Then d and β must agree in sign.

By successively forming the (hyperbolic) inner product of $Q = \alpha P + \beta A$ with A, Q, and P, show that

$$\alpha = 1/\lambda, \quad \beta = k^2(\lambda - \frac{1}{\lambda})/2$$

$$\langle P,Q \rangle = -k^2(\lambda + \frac{1}{\lambda})/2 = -k^2 \cosh(\ln \lambda)$$

Deduce that d is positive if $\lambda > 1$ and negative if $\lambda < 1$, and show that $d = k \ln \lambda$.

2. (See Exercise 1.) Let $A_\theta = (1, \cos \theta, \sin \theta)$. $[A_\theta]$ is an ideal point. Let $I = (k,0,0)$. By Theorem VII-8, the horocycle through I with center $[A_\theta]$ has equation $\langle X - I, A_\theta \rangle = 0$, or $-x_0 + x_1 \cos \theta + x_2 \sin \theta = -k$. Show that the concentric horocycle at (signed) distance d from this horocycle has equation $-x_0 + x_1 \cos \theta + x_2 \sin \theta = -k e^{-d/k}$.

3. *Intersection of horocycle with radius.* Suppose P is any point on the horocycle $\langle A,X \rangle = -1$, and let $X(s) = [\cosh(s/k)] P + [\sinh(s/k)] T$ be a distance parametrization of the horocycle's radius

through P. Let $Y(t)$ be a parametrization of the horocycle such that $Y(0) = P$ and $Y'(0) \neq O$. Prove that the horocycle and its radius intersect orthogonally, i.e., show that $<T,Y'(0)> = 0$. (Hint: Since $T \in [P,A]$, it is enough to show that $<P,Y'(0)> = <A,Y'(0)> = 0$.)

4. Let $[A]$ be one of the two ideal points on line ℓ. Prove that the plane through the origin spanned by A and vector ℓ is tangent to the light cone (along the element $[A]$). (Hint: the plane consists of all linear combinations of A and ℓ. Find the plane's intersection with the light cone.)

8. INTERSECTIONS

Suppose $<X,\ell> = 0$ and $<X,m> = 0$ are the equations of two lines on H^2, where $<\ell,\ell> = <m,m> = k^2$. Any vector of $\mathbb{R}^3$ which satisfies both of these equations, being e-orthogonal to both ℓ_{ref} and m_{ref}, must be a scalar multiple of $\ell_{ref} \times m_{ref} = -(\ell \times m)_{ref}$. Such a vector may lie either in, on, or outside the light cone. There are thus three cases to consider, according as $<(\ell \times m)_{ref},(\ell \times m)_{ref}> = <\ell \times m, \ell \times m>$ is negative, zero, or positive, and these cases will turn out to correspond, respectively, to the cases of intersecting, parallel, and divergent lines.

If $<\ell \times m, \ell \times m> < 0$, then there is a unique real number c for which $P = c(\ell \times m)_{ref}$ is on H^2. This is the case of intersecting lines: ℓ and m pass through P. (Cf. Exercise 2, Section 5.)

If $<\ell \times m, \ell \times m> = 0$, then the planes which contain the lines ℓ and m intersect in an element of the light cone, $[(\ell \times m)_{ref}]$. This is the case of parallel lines.

Lastly, if $<\ell \times m, \ell \times m> > 0$, there is a real number c, unique except for sign, for which $P = c(\ell \times m)_{ref}$ satisfies $<P,P> = k^2$. Since P lies on the single-sheeted hyperboloid $<X,X> = k^2$, we know from Section 2 that P determines a Lobachevskian line, namely the set of all X on H^2 satisfying $<X,P> = 0$. In keeping with our notational conventions, we shall refer to this line as "the line P."

Since, by hypothesis, $<\ell,P> = <m,P> = 0$, Exercise 1 of Section 5 tells us that line P is a common perpendicular to lines ℓ and m; ℓ and m are thus divergent, and line P is the axis of the associated ultra-ideal point Γ. Moreover, Γ consists of precisely those plane sections of H^2 which pass through (the origin and) point P. This is because a line q is perpendicular to the axis, line P, if and only if $<q,P> = 0$, and this is exactly the condition that the plane $<X,q> = 0$ pass through P.

INTERSECTIONS   275

Conversely, a point **P** satisfying $\langle P,P \rangle = k^2$ determines an ultra-ideal point. Its axis is the line with equation $\langle X,P \rangle = 0$, and its members are the Lobachevskian lines whose planes pass through **P**. The opposite point -**P** determines the same ultra-ideal point.

You will note that we now have two different ways of thinking of points on the single-sheeted hyperboloid $\langle X,X \rangle = k^2$. On the one hand, any vector ℓ on this hyperboloid corresponds to a Lobachevskian line, $\langle X,\ell \rangle = 0$. On the other hand (in the context of the preceding paragraph), a point **P** satisfying $\langle P,P \rangle = k^2$ represents an ultra-ideal point. (In both interpretations, we identify antipodal points.)

The relationship between these two interpretations is this: the common perpendicular to two divergent lines which meet in the ultra-ideal point **P** is the Lobachevskian line with equation $\langle X,P \rangle = 0$.

Exercises VII-8

1. *Distance between divergent lines.* Show that the perpendicular distance between two divergent lines $\langle X,\ell \rangle = 0$ and $\langle X,m \rangle = 0$ is given by

$$\cosh^2 \frac{d}{k} = \frac{\langle \ell,m \rangle^2}{\langle \ell,\ell \rangle \langle m,m \rangle}$$

(Hint: let the common perpendicular, $\langle X,p \rangle = 0$, meet $\langle X,\ell \rangle = 0$ in **L** and $\langle X,m \rangle = 0$ in **M**. Then the distance d from **L** to **M** is the distance between the lines. Since $\langle L,\ell \rangle = \langle L,p \rangle = \langle M,m \rangle = \langle M,p \rangle = 0$, **L** and **M** are proportional to $(\ell \times p)_{ref}$ and $(m \times p)_{ref}$, respectively. Use (32) to express $\cosh^2 d/k$ in terms of ℓ and **m**.)

9. EQUIDISTANT CURVES

Let $X(s) = (\cosh \frac{s}{k})P + (\sinh \frac{s}{k})T$ be a distance parametrization of the Lobachevskian line ℓ. As usual, we assume $\langle T,T \rangle = \langle \ell,\ell \rangle = k^2$ and $\langle P,T \rangle = 0$. Let

$$Y_s(u) = (\cosh \frac{u}{k})X(s) + (\sinh \frac{u}{k})\ell$$

and check that $\langle Y_s,Y_s \rangle = -k^2$ for all s, u. Because $\langle T,\ell \rangle = 0$, $Y_s(u)$

is, for fixed s, a distance parametrization of the Lobachevskian line perpendicular to ℓ at the point $X(s)$. Therefore, for fixed $u = d > 0$ (and varying s), the mapping $s \to Y_s(d)$ is (one branch of) the equidistant curve with axis ℓ and distance d. The other branch is obtained by replacing ℓ by $-\ell$ in the formula for Y_s.

Note that $<Y_s(d),\ell> = k^2 \sinh \frac{d}{k}$, and so as s varies, $Y_s(d)$ traces out the plane section of H^2 consisting of all X satisfying $<X,\ell> = k^2 \sinh(d/k)$. Note also that the plane of this curve is parallel (in the Euclidean sense) to the plane $<X,\ell> = 0$. (Both planes meet H^2 in branches of hyperbolas.)

Theorem VII-9

A branch of an equidistant curve is represented on H^2 by a plane section whose plane is parallel (in the Euclidean sense) to the plane of the curve's axis. If $<\ell,\ell> = k^2$ and $d > 0$, then the set of all $X \in H^2$ satisfying

$$<X,\ell> = k^2 \sinh \frac{d}{k}$$

constitutes one branch of the equidistant curve with axis ℓ and distance d. The other branch is given by $<X,-\ell> = k^2 \sinh(d/k)$.

In terms of homogeneous coordinates, the equidistant curve equation takes the form

$$\frac{<X,\ell>^2}{-<X,X><\ell,\ell>} = \sinh^2 \frac{d}{k} \tag{47}$$

and this includes both branches.

Corollary VII-10

The distance d from a point P of H^2 to the line ℓ (where $<\ell,\ell> = k^2$) is given by

$$\sinh \frac{d}{k} = \frac{1}{k^2} |<P,\ell>|$$

In homogeneous coordinates, the distance is given by

EQUIDISTANT CURVES

$$\sinh^2 \frac{d}{k} = \frac{\langle P,\ell \rangle^2}{-\langle P,P \rangle \langle \ell,\ell \rangle}$$

Exercises VII-9

1. *Spherical and elliptic geometry.* Let R be a positive real number and denote by S the set of all $X \in \mathbb{R}^3$ satisfying $X \cdot X = R^2$ (a sphere of radius R). In *spherical geometry,* the "lines" are great circles, i.e., sections of S by planes through the origin.

    (a) If $X, Y \in S$, the distance d between X and Y is the length of the shortest great circle arc joining X to Y. Show

$$\cos \frac{d}{R} = \frac{1}{R^2} (X \cdot Y)$$

Compare this with (33). (Hint: relate d to the angle subtended by the arc from X to Y.)

Since a great circle lies on a plane through the origin, a line in spherical geometry has an equation of the form $X \cdot P = 0$, where P is fixed. If we require $P \cdot P = R^2$, then P is unique up to sign.

A model for *elliptic geometry* is obtained from S by identifying (considering as the same) each $X \in S$ and its opposite, $-X$. If $X \cdot P = 0$ is the equation of a line of the elliptic plane, where $P \cdot P = R^2$, then P is called the *pole* of the line, and the line is called the *polar* of P. (There is only one pole, since P and $-P$ represent the same point.)

    (b) Show geometrically that in the elliptic plane, the perpendicular distance d between a point P and a line $X \cdot Q = 0$ is given by

$$\sin \frac{d}{R} = \frac{1}{R^2} |P \cdot Q|$$

Compare with Cor. VII-10. (Hint: Draw a sketch; $d = \pi R/2 - PQ$.)

    (c) Show geometrically that the angle θ between the lines $X \cdot P = 0$ and $X \cdot Q = 0$ is $1/R$ times the distance between their poles, and deduce that $\cos \theta = (P \cdot Q)/R^2$, or

$$\cos^2 \theta = \frac{(P \cdot Q)^2}{(P \cdot P)(Q \cdot Q)}$$

Compare the formula of Exercise 1, Section 5. Dually, the distance between two points is R times the angle between their polars.

The similarity between the formulas for distance and angle is less mysterious in elliptic geometry because of the pole-polar correspondence. Such a correspondence between points and lines is possible for H^2 only when we allow the coordinates to assume complex values. This approach is beyond the scope of this text.

2. (*Klein-Beltrami Disk*) For each point $P = (p_0, p_1, p_2)$ on H^2, the straight line [P] meets the plane $x_0 = 1$ in a unique point $(1, u, v)$.

(a) Show that $u = p_1/p_0$, $v = p_2/p_0$, and $u^2 + v^2 < 1$.

(b) Prove that the correspondence $P \to (u,v)$ is a one-to-one mapping of H^2 onto the open unit disk $D = \{(u,v) \in \mathbb{R}^2 | u^2 + v^2 < 1\}$.

(c) Show that lines of H^2 correspond to chords of D.

(d) Define the distance d between points (u,v) and (u', v') of D to be the distance between the corresponding points P and P' of H^2. Show that

$$\cosh \frac{d}{k} = \frac{1 - uu' - vv'}{(1 - u^2 - v^2)^{1/2}(1 - u'^2 - v'^2)^{1/2}}$$

VIII

LOBACHEVSKIAN GEOMETRY AND PHYSICAL SPACE

> ... *the axioms of geometry ... are only definitions in disguise. What then are we to think of the question: Is Euclidean Geometry true? It has no meaning. We might as well ask if the metric system is true and if the old weights and measures are false; if Cartesian coordinates are true and polar coordinates false. One geometry cannot be more true than another: it can only be more convenient.*
> —Henri Poinca_é

INTRODUCTION

What is the geometry of the physical universe? Surely this question cannot be decided on the basis of logic alone, for as we know now, there are geometries other than Euclid's, just as consistent, and there is no a priori reason for preferring one over any other. Therefore, with Gauss and Lobachevsky, we would shift the burden to observation and experiment.

However, here there is great difficulty. In order to apply an abstract geometric theory to the study of physical space, we must decide which physical entities are to represent the primitive terms of the theory: point, line distance, etc. Although we all have a fairly good intuitive notion of what point and line are, it is not at all clear how to define a physical point or a physical line.

Indeed, there might not exist in nature absolute things that correspond to point, line, distance, etc., and even if there did, they might be beyond our power to quantify or measure. Ultimately, all our knowledge of the physical world rests on the behavior of material measuring devices, and so all our inferences concerning the geometry of nature inevitably involve the interactions of matter. It is impossible to separate applied geometry from physical laws.

At best, we can give only approximate representations of the primitive terms, and see if our measured data concerning these representations are in approximate conformity with one particular geometry or another. We may, for example, think of a point as the limit approached by a small volume of space as its dimensions approach zero. Perhaps this idea was behind Euclid's definition of a point as "that which has no part." A line is perhaps represented by the path of a light ray.

In this chapter we shall engage in some speculation concerning this introduction's opening question. Admittedly, the approach will be somewhat naive, as is unavoidable without the theoretical background of modern astronomy and relativity theory.

1. DEFECTS AND THE PARALLAX OF STARS

In Lobachevskian geometry, the area of a triangle is proportional to its defect. The fact that the formulas of Lobachevskian geometry approach their Euclidean counterparts as the dimensions of figures approach zero can be used to show that the constant of proportionality must be k^2 if defects are given in radians. Therefore, the area of a triangle with defect δ is

$$A = k^2 \delta \tag{48}$$

(Since δ cannot exceed π, $k^2 \pi$ is an upper bound for the areas of triangles, as was known to Gauss.) Perhaps the first type of experiment that comes to mind for probing the geometry of space is the measurement of angle sums in triangles. If we could find a triangle with positive defect, we would establish Lobachevskian geometry for space and also, by (48), determine the space constant k as well, at least in principal.

Gauss was probably one of the first for whom the question of whether Euclid's postulate is "true" was of actual physical signifi-

DEFECTS AND THE PARALLAX OF STARS

cance. His measurement of the angle sum in a triangle formed by three mountain peaks may have had as a secondary purpose the testing of Euclid's postulate. Since the deviation from 180° was no greater than the inevitable errors inherent in imperfect measuring processes, nothing was settled (cf. Section IV-8).

We can see why this could have been expected if we pass to a much larger triangle, say that formed by the centers of the sun, the earth, and Mars. This is more than a trillion (10^{12}) times larger than the triangle Gauss measured. Accordingly, even if this larger triangle had a defect as large as one degree, the triangle of Gauss would have a defect on the order of 0.000000000001°, too minute for our instruments to measure, even today.

Lobachevsky considered triangles far larger in area than any contained within our solar system. He examined the *parallax* of stars, the annual oscillation in a star's apparent position because of the earth's motion around the sun.

In Figure VIII-1a, A and B are two positions of the earth in its orbit six months apart. S is a relatively nearby star. As the earth moves in its orbit, the direction of the earth-to-star line of sight changes, with the result that the star appears to move in a small circle in the heavens during the course of a year. This circle is smaller for a more remote star (S' in Fig. VIII-1), and this variation in size is used to measure stellar distances. (In practice, parallax is measured by comparing the apparent position of a star with that of some reference star known to be much more distant.)

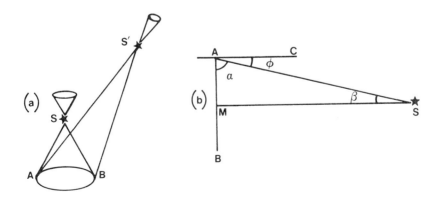

Figure VIII - 1.

For convenience, assume S lies on a perpendicular bisector of AB (Fig. VIII-1b). Let $\alpha = \angle MAS$, $\beta = \angle ASM$, and $\overleftrightarrow{AC} \perp \overleftrightarrow{AB}$. The angle $\phi = \angle CAS$ is called the *parallax angle* of the star. Since

$$\mathscr{D}(ASM) = \pi - (\frac{\pi}{2} + \alpha + \beta) < \frac{\pi}{2} - \alpha = \phi$$

the defect is smaller than the parallax angle. Even for the nearest stars, parallax angles are very tiny, e.g., about 0.783 seconds of arc for Alpha Centauri. (This is the angle subtended by a dime at a distance of 5 km.) The defect of $\triangle ASM$, if non-zero, must then itself be extremely small. From (48), it follows then that if space is Lobachevskian, the space constant must be quite large. In fact, on the basis of some rough calculations (Exercise 2), we can show that

$$k > \frac{R}{2 \tan(\phi/2)}$$

where $R = AM$ is the mean distance from the earth to the sun. For Alpha Centauri, this yields $k > (2.63 \times 10^5)R$, or about 4.4 light years; and more distant stars would give larger lower bounds on k. (Actually, parallax measurements are practicable only up to distances on the order of 100 light years.)

Looking at the matter another way, for any star S, $\alpha = \pi/2 - \phi$ must be less than the angle of parallelism, $\Pi(R)$, so that, for any star,

$$\phi > \frac{\pi}{2} - \Pi(R)$$

Accordingly, if space is Lobachevskian, there exists a lower bound for the parallax of stars. No such lower bound has ever been uncovered, nor has any stellar triangle been proven to have a positive defect. Of course, this does not prove that space is Euclidean. We can conclude only that if space is Lobachevskian, then the space constant k is exceedingly large.

It should be pointed out that we could *never* prove space is Euclidean in this manner. The failure to measure a deviation from 180° larger than the inaccuracies inherent in making measurements could mean either that light rays conform to Euclidean geometry or

DEFECTS AND THE PARALLAX OF STARS

that we simply are looking at triangles that are too small compared to k. On the other hand, if space really is Lobachevskian, it is conceivable that sufficiently precise measurements might eventually show it.

Of course it comes as no surprise that the results of optical experiments within our solar system or within our local star group are consistant with Euclidean geometry. After all, our science and technology have succeeded reasonably well with Euclidean geometry for some time. However, our faith in this geometry is based on experience that is really very limited when gauged by the vastness of the universe. Is it not possible that on a very large scale or in some distant region our measurements would conform to a different kind of geometry?

Exercises VIII-1

1. Use the parallax of Proxima Centauri ($\phi \approx 0.783''$) to estimate the star's distance from the sun. (Do not forget to convert to radians. $R \approx 1.496 \times 10^{13}$ cm. The actual distance is about 4.29 light-years. A light-year is approximately 9.46×10^{17} cm.)

2. Let R = AM in Figure VIII-1b. Show that $R/k < 2 \tan(\phi/2)$, or $k > R/[2 \tan(\phi/2)]$. Hint: you will need the following:

$$\tan \frac{\Pi(R)}{2} = e^{-R/k}$$

$$\ln(\frac{1+x}{1-x}) \approx 2x \quad \text{for x small}$$

$$\tan(\frac{\pi}{4} - \theta) = \frac{1 - \tan \theta}{1 + \tan \theta}$$

[$\Pi(R)$ is the angle of parallelism for R.]

3. Verify the estimate $k > 2.63 \times 10^5$ R, based upon the parallax angle $\phi \approx 0.783''$ of Proxima Centauri.

4. For the star Vega, the parallax angle ϕ is about $0.08''$. What lower bound does this imply for k if space is Lobachevskian?

2. THE FINITE CURVED UNIVERSE

In all fairness, we should give equal time to the possibility that observations on a large scale might show that space is more accurately represented by spherical or elliptic geometry, so that triangles would have angular excesses rather than defects. This raises the intriguing question of whether the universe might be finite in volume, as the surface of a sphere is finite in area.

To many laymen the notion of a finite universe seems paradoxical. If the universe is finite, they argue, then there must be something outside, something "beyond." But the word "universe," by definition, means "all," and so this "outside" must be a part of the universe too, and therefore inside.

This apparent contradiction stems both from a confusion of the two concepts "with boundary" and finite (cf. Section III-3), and from an incorrect generalization of a spherical surface to three dimensions. If the universe is finite, then space is the 3-dimensional analogue of the surface of a ball, not the interior of a ball. Just as a ball's surface (in mathematical jargon, the 2-sphere) has no edge, so too a 3-dimensional spherical space (the 3-sphere), has no boundary. The concepts *finite* and *having boundary* are independent. This is shown clearly for two dimensions in Table VIII-1.

The conceptual difficulties are no doubt due to the near impossibility of visualizing objects in higher dimensions. When we think of a 2-sphere, we naturally picture the 3-dimensional ball that it encloses. Indeed we find it virtually impossible to picture a 2-sphere except as imbedded in E^3 (Euclidean 3-dimensional space) as the boundary of a ball. We are unable to visualize a 3-sphere because we cannot picture an enclosed 4-dimensional ball in E^4. However, the 3-sphere, like the 2-sphere, is a mathematical object that is definable without reference to a higher-dimensional containing space. How this is possible is made clear in courses on differential geometry or manifold theory. [The 3-sphere may be represented as the set of all points (x,y,z,w) in $\mathbb{R}^4$ such that $x^2 + y^2 + z^2 + w^2 = 1$. However, this representation requires a space of four dimensions.]

Table VIII - 1

	Plane	Sphere	Closed $\frac{1}{2}$-Plane	Closed Disk
Finite	No	Yes	No	Yes
With Boundary	No	No	Yes	Yes

THE FINITE CURVED UNIVERSE

The boundary of any object, of whatever dimension, is an object of one lower dimension. The boundary of a 3-dimensional ball is a 2-sphere, but the 2-sphere itself has no boundary. Analogously, the boundary of a 4-dimensional ball is a 3-sphere, but the 3-sphere has no boundary (and has finite volume). If the universe is spatially finite, it could conceivably be something like a 3-sphere.

It is difficult for us to imagine such a "curved" universe, but we can gain something of an inutitive or qualitative appreciation for curvature by imagining a 2-dimensional universe, as in E. A. Abbott's fantasy, *Flatland* [1], and drawing comparisons with the high-dimensional situation.

Suppose the inhabitants of Flatland measure the area $A(r)$ of a disk of radius r, for various values of r. (In this context a disk is the set of all points within distance r from a given point.) If their space is flat (zero curvature), they would find that $A(r)$ grows in proportion to r^2, since, in the plane, $A = \pi r^2$.

Suppose, however, their universe is the surface of a sphere of radius R. The radius r of a disk is now measured along an arc of a great circle (Fig. VIII-2). Now $A(r)$ grows at a proportionately slower rate than r^2.

For example, if r is equal to one-quarter the circumference of a great circle, then $A(r)$ is the area of a hemisphere. If we double r, then $A(r)$ becomes the area of the entire sphere. Doubling the radius only doubled the area, whereas in the plane the area would have been quadrupled. In general, disks on the sphere have less area than their counterparts of the same radius in the plane (see Exercise 1). This is

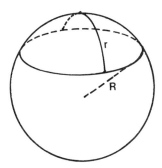

Figure VIII - 2.

why a piece of orange peel, when pressed flat onto a table, necessarily stretches and tears. (It does not have enough area to fill a disk of the same radius in the plane.) A surface on which A(r) grows more slowly than πr^2 will be said to have *positive curvature*.

Dually, a surface on which A(r) grows more rapidly then πr^2 will be said to have *negative curvature*. Such surfaces are "saddle-shaped," as for example, a hyperbolic paraboloid. Historically, a most important example is Beltrami's pseudosphere, the first model for Lobachevskian geometry (cf. Section VI-3).

In courses on differential geometry, curvature is defined as a certain function that assigns a number to each point of a surface, and in general curvature varies from point to point. The plane, the sphere, and the pseudosphere are special in that the curvature is constant on each of them.

The three possible cases of 2-dimensional constant curvature geometries—curvature zero, curvature positive, and curvature negative—are, respectively, Euclidean or parabolic, spherical or elliptic, and Lobachevskian or hyperbolic. *Elliptic* derives from a Greek word meaning deficit or falling short: in elliptic geometry there are no parallels. *Hyperbolic* comes from a Greek root signifying excess or overshoot: in hyperbolic geometry there are too many parallels. *Parabolic* is from a Greek word meaning to throw beside or compare. This is the intermediate case of unique parallels.

Drawing an analogy with the 2-dimensional case, we might expect that curvature in three dimensions might be detectable by comparing the growth in the volume of a ball of radius r with $(4/3)\pi r^3$ (the volume of a ball in Euclidean space). We might, for example, try to count the galaxies (which, on a very large scale, appear to be uniformly distributed) that lie within distance r of our own, and see how this varies with r. Perhaps there is some overall average non-zero curvature to the universe—either positive or negative—even though on the scale of our solar system, space appears flat.

Before leaving these speculations, it should be pointed out that one essential factor, time, has been completely ignored so far. When we gaze into the distant reaches of the heavens, we also look back in time, since the starlight we are seeing left its origin eons ago. We cannot ignore the possibility that the geometry of the universe is dynamic and changing, rather than static. (Twentieth century astronomy has indeed established that this is actually the case.)

THE FINITE CURVED UNIVERSE

Exercises VIII-2

1. (a) Show, using calculus, that the surface area A(r) of a disk of radius r in spherical geometry (i.e., the area of the spherical segment shown in Fig. VIII-2) is

$$A(r) = 2\pi R^2 \left(1 - \cos\frac{r}{R}\right) = 4\pi R^2 \sin^2 \frac{r}{2R}$$

(b) Show that $A(r) < \pi r^2$ for all $r > 0$, and that

$$\lim_{r/R \to 0} \frac{A(r)}{\pi r^2} = 1$$

2. By replacing R by ik in the formulas for A(r) derived in Exercise 1, find analogous formulas for the area A(r) of a disk of radius r in Lobachevskian geometry. Show that A(r) is now greater than πr^2.

3. In 3-dimensional Lobachevskian geometry, the surface area S of a sphere of radius r is

$$S = 4\pi k^2 \sinh^2 \frac{r}{k}$$

(a) Show that if $r/k \ll 1$, then $S \approx 4\pi r^2$, the surface area of a sphere in Euclidean geometry. (b) Show however that as r/k increases, S grows much faster than $4\pi r^2$. For example, estimate $S/(4\pi r^2)$ when r = 5k. (c) If physical space were Lobachevskian, the observed brightness of a star would decrease in proportion to $1/\sinh^2 (r/k)$, where r is the observer's distance from the star (instead of in proportion to $1/r^2$ under the assumption of Euclidean geometry.) This is because a spreading spherical light wave carrying, say, one unit of light energy spreads that energy evenly over a spherical surface of area S when the wave front has radius r. Therefore the observed brightness (energy received per unit area) is 1/S. If space were Lobachevskian, would the stars be nearer or farther away than when their distances are computed from measurements of observed brightness using Euclidean geometry?

3. PHILOSOPHICAL OBJECTIONS: TRUTH OR CONVENIENCE

Like Gauss, Henri Poincaré (1854-1912) was a "universalist," whose contributions spanned practically all branches of mathematics, including complex function theory, analysis, topology, and probability theory. He contributed also to philosophy, physics, and cosmology, and was one of the pioneers of relativity theory.

Since all measurements involve physical and geometric assumptions, Poincaré argued that it was meaningless to ask whether space is Euclidean or non-Euclidean. If we measure a triangle and detect a defect, we could say space is non-Euclidean, but we could equally well claim that space is Euclidean and that light rays do not always follow straight line paths. With equal impunity we could place the blame on geometry or on physics.

Poincaré emphasized his point by means of an imaginary universe Σ occupying the interior of a sphere of radius R in E^3, and in which the following laws hold:

1. At any point P of Σ, the temperature T is given by

   $$T(P) = k(R^2 - r^2)$$

   where k is a positive constant, and r is the Euclidean distance from the sphere's center to P.
2. The linear dimensions of any material body vary directly with the temperature of the body's locality.
3. All material bodies immediately assume the temperature of their localities.

Because of (1) and (2), an inhabitant of Σ shrinks as he approaches the boundary of Σ, so that he would need an infinite number of steps to reach the boundary. Moreover, he cannot detect his shrinkage because his measuring sticks shrink right along with him; and he cannot feel the changes in temperature because of (3). Hence to him, his universe appears to be infinite.

Now it can be shown that the paths of shortest length—the so-called *geodesics*—joining pairs of points in Σ are not straight lines, as they would be in the absence of shrinkage, but curve inward toward the center of Σ (Fig. VIII-3). They are actually circular arcs meeting the boundary sphere orthogonally.

PHILOSOPHICAL OBJECTIONS

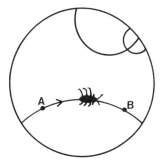

Figure VIII - 3. Bug's right legs shorter than left because of laws (1), (2), and (3). Bug thinks it is following straight line, but unequal legs cause it to trace circular arc $\widehat{AB}$.

Let us assume one additional law:

4. Light rays travel along geodesics.

This will make the geodesics "look" straight to the inhabitants of Σ. (For instance, the light rays from a collection of light bulbs strung out along a geodesic will all follow this geodesic and so enter an observer's eyes from the same direction.)

Since Lobachevsky's postulate holds in the geodesic geometry of Σ (see Fig. VIII-3), the inhabitants would believe they are in a non-Euclidean world. Here we have a region of ordinary Euclidean space which, because of strange (and undetectable) physical laws, appears non-Euclidean to its inhabitants.

If we were suddenly transported to Σ, we could adopt the viewpoint of its lifelong inhabitants and assert the geometry of the world is Lobachevskian; or we might cling to our Euclidean beliefs and declare that the physical properties of space are such that light rays bend and things vary in size. Both viewpoints are equally correct.

To Poincaré, then, it was meaningless to ask which is the *true* geometry of space. The question becomes instead, "Which geometry is the most *convenient* for describing the physical world?" For terrestrial engineering and most solar system astronomy, Euclidean geometry is the more convenient because of its greater simplicity. But on a larger scale, and in the realm of strong gravitational fields, the non-Euclidean 4-dimensional spacetime geometry of Einstein has proved far more convenient.

APPENDIX A

DEFINITIONS, POSTULATES, PROPOSITIONS OF EUCLID, BOOK I*

DEFINITIONS
1. A **point** is that which has no part.
2. A **line** is breadthless length.
3. The extremities of a line are points.
4. A **straight line** is a line which lies evenly with the points on itself.
5. A **surface** is that which has length and breadth only.
6. The extremities of a surface are lines.
7. A **plane surface** is a surface which lies evenly with the straight lines on itself.

---
*Reprinted with permission from *The Thirteen Books of Euclid's Elements*, translated by Sir Thomas L. Heath, 2nd ed., Dover Publications, New York, 1956.

8. A **plane angle** is the inclination to one another of two lines in a plane which meet one another and do not lie in a straight line.

9. And when the lines containing the angle are straight, the angle is called **rectilineal**.

10. When a straight line set up on a straight line makes the adjacent angles equal to one another, each of the equal angles is right, and the straight line standing on the other is called a **perpendicular** to that on which it stands.

11. An **obtuse angle** is an angle greater than a right angle.

12. An **acute angle** is an angle less than a right angle.

13. A **boundary** is that which is an extremity of anything.

14. A **figure** is that which is contained by any boundary or boundaries.

15. A **circle** is a plane figure contained by one line such that all the straight lines falling upon it from one point among those lying within the figure are equal to one another;

16. And the point is called the **centre** of the circle.

17. A **diameter** of the circle is any straight line drawn through the centre and terminated in both directions by the circumference of the circle, and such a straight line also bisects the circle.

18. A **semicircle** is the figure contained by the diameter and the circumference cut off by it. And the centre of the semicircle is the same as that of the circle.

19. **Rectilineal figures** are those which are contained by straight lines, **trilateral** figures being those contained by three, **quadrilateral** those contained by four, and **multilateral** those contained by more than four straight lines.

20. Of trilateral figures, an **equilateral triangle** is that which has its three sides equal, an **isosceles triangle** that which has two of its sides alone equal, and a **scalene triangle** that which has its three sides unequal.

21. Further, of trilateral figures, a **right-angled triangle** is that which has a right angle, an **obtuse-angled triangle** that which has an obtuse angle, and an **acute-angled triangle** that which has its three angles acute.

22. Of quadrilateral figures, a **square** is that which is both equilateral and right-angled; an **oblong** that which is right-angled but not equilateral; a **rhombus** that which is equilateral but not right-angles; and a **rhomboid** that which has its opposite sides and angles

DEFINITIONS

equal to one another but is neither equilateral nor right-angled. And let quadrilaterals other than these be called **trapezia**.

23. **Parallel** straight lines are straight lines which, being in the same plane and being produced indefinitely in both directions, do not meet one another in either direction.

THE POSTULATES

Let the following be postulated:

1. To draw a straight line from any point to any point.
2. To produce a finite straight line continuously in a straight line.
3. To describe a circle with any centre and distance.
4. That all right angles are equal to one another.
5. That, if a straight line falling on two straight lines make the interior angles on the same side less than two right angles, the two straight lines, if produced indefinitely, meet on that side on which are the angles less than the two right angles.

THE COMMON NOTIONS

1. Things which are equal to the same thing are also equal to one another.
2. If equals be added to equals, the wholes are equal.
3. If equals be subtracted from equals, the remainders are equal.
4. Things which coincide with one another are equal to one another.
5. The whole is greater than the part.

(The Common Notions were probably considered to be fundamental rules of inference that were common to all deductive disciplines, as distinct from the Postulates, which were peculiar to geometry. By analogy, in a rigorous presentation of a modern mathematical theory, it is either explicitly stated or automatically assumed that standard principles of logical inference, Zermelo-Frankel set theory, etc., are followed.)

THE PROPOSITIONS OF BOOK I

1. On a given finite straight line to construct an equilateral triangle.
2. To place at a given point (as an extremity) a straight line equal to a given straight line.

3. Given two unequal straight lines, to cut off from the greater a straight line equal to the less.

4. If two triangles have the two sides equal to two sides respectively, and have the angles contained by the equal straight lines equal, they will also have the base equal to the base, the triangle will be equal to the triangle, and the remaining angles will be equal to the remaining angles respectively, namely those which the equal sides subtend.

5. In isosceles triangles the angles at the base are equal to one another, and, if the equal straight lines be produced further, the angles under the base will be equal to one another.

6. If in a triangle two angles be equal to one another, the sides which subtend the equal angles will also be equal to one another.

7. Given two straight lines constructed on a straight line (from its extremities) and meeting in a point, there cannot be constructed on the same straight line (from its extremities), and on the same side of it, two other straight lines meeting in another point and equal to the former two respectively, namely each to that which has the same extremity with it.

8. If two triangles have the two sides equal to two sides respectively, and have also the base equal to the base, they will also have the angles equal which are contained by the equal straight lines.

9. To bisect a given rectilineal angle.

10. To bisect a given finite straight line.

11. To draw a straight line at right angles to a given straight line from a given point on it.

12. To a given infinite straight line, from a given point which is not on it, to draw a perpendicular straight line.

13. If a straight line set up on a straight line make angles, it will make either two right angles or angles equal to two right angles.

14. If with any straight line, and at a point on it, two straight lines not lying on the same side make the adjacent angles equal to two right angles, the two straight lines will be in a straight line with one another.

15. If two straight lines cut one another, they make the vertical angles equal to one another.

16. In any triangle, if one of the sides be produced, the exterior angle is greater than either of the interior and opposite angles.

17. In any triangle two angles taken together in any manner are less than two right angles.

18. In any triangle the greater side subtends the greater angle.

19. In any triangle the greater angle is subtended by the greater side.

20. In any triangle two sides taken together in any manner are greater than the remaining one.

21. If on one of the sides of a triangle, from its extremities, there be constructed two straight lines meeting within the triangle, the straight lines so constructed will be less than the remaining two sides of the triangle, but will contain a greater angle.

22. Out of three straight lines, which are equal to three given straight lines, to construct a triangle: thus it is necessary that two of the straight lines taken together in any manner should be greater than the remaining one.

23. On a given straight line and at a point on it to construct a rectilineal angle equal to a given rectilineal angle.

24. If two triangles have the two sides equal to two sides respectively, but have the one of the angles contained by the equal straight lines greater than the other, they will also have the base greater than the base.

25. If two triangles have the two sides equal to two sides respectively, but have the base greater than the base, they will also have the one of the angles contained by the equal straight lines greater than the other.

26. If two triangles have the two angles equal to two angles respectively, and one side equal to one side, namely, either the side adjoining the equal angles, or that subtending one of the equal angles, they will also have the remaining sides equal to the remaining sides and the remaining angle to the remaining angle.

27. If a straight line falling on two straight lines make the alternate angles equal to one another, the straight lines will be parallel to one another.

28. If a straight line falling on two straight lines make the exterior angle equal to the interior and opposite angle on the same side, or the interior angles on the same side equal to two right angles, the straight lines will be parallel to one another.

29. A straight line falling on parallel straight lines makes the

alternate angles equal to one another, the exterior angle equal to the interior and opposite angle, and the interior angles on the same side equal to two right angles.

30. Straight lines parallel to the same straight line are also parallel to one another.

31. Through a given point to draw a straight line parallel to a given straight line.

32. In any triangle, if one of the sides be produced, the exterior angle is equal to the two interior and opposite angles, and the three interior angles of the triangle are equal to two right angles.

33. The straight lines joining equal and parallel straight lines (at the extremities which are) in the same directions (respectively) are themselves also equal and parallel.

34. In parallelogrammic areas the opposite sides and angles are equal to one another, and the diameter bisects the areas.

35. Parallelograms which are on the same base and in the same parallels are equal to one another.

36. Parallelograms which are on equal bases and in the same parallels are equal to one another.

37. Triangles which are on the same base and in the same parallels are equal to one another.

38. Triangles which are on equal bases and in the same parallels are equal to one another.

39. Equal triangles which are on the same base and on the same side are also in the same parallels.

40. Equal triangles which are on equal bases and on the same side are also in the same parallels.

41. If a parallelogram have the same base with a triangle and be in the same parallels, the parallelogram is double of the triangle.

42. To construct, in a given rectilineal angle, a parallelogram equal to a given triangle.

43. In any parallelogram the complements of the parallelograms about the diameter are equal to one another.

44. To a given straight line to apply, in a given rectilineal angle, a parallelogram equal to a given triangle.

45. To construct, in a given rectilineal angle, a parallelogram equal to a given rectilineal figure.

46. On a given straight line to describe a square.

47. In right-angled triangles the square on the side subtending

the right angle is equal to the squares on the sides containing the right angle.

48. If in a triangle the square on one of the sides be equal to the squares on the remaining two sides of the triangle, the angle contained by the remaining two sides of the triangle is right.

APPENDIX B

HILBERT'S POSTULATES

The undefined entities are *points* (designated by A, B, C, etc.), *lines* (designated a, b, c, etc.), and *planes* (designated α, β, γ, etc.). There are in addition the relations *lies on, between,* and *congruent* (whose exact meanings will be supplied by the postulates).

The following five groups of axioms constitute Hilbert's postulates for *space* (i.e., solid Euclidean) *geometry*.* Hilbert's postulates for *plane geometry* are obtained by omitting all phrases and sentences which appear below enclosed within brackets. When Hilbert says two, three, or more points, lines, or planes, he always intends that these are distinct points, lines, or planes. If A is a point

---

*Reprinted from *Foundations of Geometry* by David Hilbert by permission of The Open Court Publishing Company, La Salle, Illinois. ©1971 by The Open Court Publishing Company, La Salle, Illinois. Translated by Leo Unger from the 10th German edition.

and a is a line, then in place of "A lies on a," we will often substitute "a contains A," "a passes through A," "A is on a," etc. Two lines containing A are said to "intersect in A," or "have A in common." Similar terminological substitutions will be used for incidence relations involving planes.

GROUP I: THE INCIDENCE AXIOMS

I,1. For every two points A, B there exists a line a that contains each of the points A, B.

I,2. For every two points A, B there exists no more than one line that contains each of the points A, B.

I,3. There exist at least two points on a line. There exist at least three points that do not lie on a line.

[I,4. For any three points A, B, C that do not lie on the same line there exists a plane α that contains each of the three points A, B, C. For every plane there exists a point which it contains.]

[I,5. For any three points A, B, C that do not lie on one and the same line there exists no more than one plane that contains each of the three points A, B, C.]

[I,6. If two points A, B of a line a lie in a plane α then every point of a lies in the plane α.]

[I,7. If two planes α, β have a point A in common then they have at least one more point B in common.]

[I,8. There exist at least four points which do not lie in a plane.]

If A and B are two distinct points of a line, we will often denote this line by $\overleftrightarrow{AB}$.

GROUP II: THE ORDER AXIOMS

II,1. If a point B lies between a point A and a point C then the points A, B, C are three distinct points of a line, and B also lies between C and A.

II,2. For two points A and C, there always exists at least one point B on the line $\overleftrightarrow{AC}$ such that C lies between A and B.

II,3. Of any three points on a line there exists no more than one that lies between the other two.

Definition B-1

Given two distinct points A and B, the segment $\overline{AB}$ consists of the points A and B together with all points which are between A and B.

GROUP II: THE ORDER AXIOMS

Definition B-2

Given three distinct points A, B, and C not all on the same line, △ABC is the union of the three segments $\overline{AB}$, $\overline{BC}$, and $\overline{CA}$.

II,4. Let A, B, C be three points that do not lie on a line and let a be a line [in the plane ABC] which does not meet any of the points A, B, C. If the line a passes through a point of the segment $\overline{AB}$, it also passes through a point of segment $\overline{AC}$, or through a point of segment $\overline{BC}$.

Let us use the notation (ABC) to indicate that B is between A and C. The following are consequences of the axioms in Groups I and II. Other consequences are deduced in the exercises of Section III-3.

Theorem B-3

Given distinct points A and C, there exists a point D such that (ADC) (Fig. B-1).

Proof: By I,3, let E be a point not on $\overleftrightarrow{AC}$. By II,2, there is a point F such that (AEF), and F ≠ C by I,2. Again by II,2, there is a point G such that (FCG), and by II,4, line $\overleftrightarrow{EG}$, which cannot pass through A, C, or F (why not?) but meets $\overline{AF}$, must meet $\overline{AC}$ or $\overline{CF}$ in a point D. But $\overleftrightarrow{EG}$ cannot meet $\overline{CF}$, for this would violate I,2. (G is not on $\overline{FC}$ by II,3.) Hence D is on $\overline{AC}$, or (ADC).

Theorem B-4

Of any three distinct points A, B, C on a line, at least one of them is between the other two (Fig. B-2).

Proof. Suppose (BAC) and (ACB) both do not hold. We shall show (ABC). By II-2, there exist points D and G such that (BDG). By II,4 (applied to △BCG), $\overline{AD}$ meets $\overline{CG}$ in E: (CEG). Similarly, $\overline{CD}$ meets $\overline{AG}$ in F: (AFG). Applying II,4 to △AEG (since $\overleftrightarrow{CF}$ meets $\overline{AG}$), $\overleftrightarrow{CF}$ must meet $\overline{AE}$ in a point between A and E. Because CF

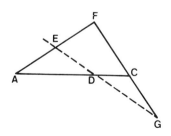

Figure B - 1.

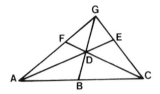

Figure B - 2.

already meets $\overleftrightarrow{AE}$ in D and can meet $\overleftrightarrow{AE}$ in no other point, then D is this point and we must have (ADE). Similarly, by II,4 applied to △AEC, since $\overleftrightarrow{GB}$ meets $\overline{AE}$, then $\overleftrightarrow{GB}$ must meet $\overline{AC}$ in some point, which must be B. Therefore (ABC).

Theorem B-5

Let A, B, C, and D be distinct points of a line. If (ABC) and (BCD), then (ABD) and (ACD) (Fig. B-3).

Proof. Let E and F be such that E is not on the line containing the four points, and (CEF). The remainder of the proof is a repeated application of Pasch's Axiom (II,4).

From △BCF, $\overleftrightarrow{AE}$ meets $\overline{BF}$ in G: (BGF).
From △AEC, $\overleftrightarrow{BF}$ meets $\overline{AE}$ in G: (AGE).
From △GBD, $\overleftrightarrow{CF}$ meets $\overline{GD}$ in H: (GHD).
From △AGD, $\overleftrightarrow{EH}$ meets $\overline{AD}$ [since (AEG) is false] in a point which must be C: (ACD).

We have thus proved that

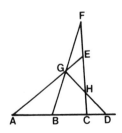

Figure B - 3.

GROUP II: THE ORDER AXIOMS

(ABC), (BCD) imply (ACD)

Now make the following permutation of the letters A, B, C, D:

A B C D
↓ ↓ ↓ ↓
D C B A

With this relabeling, the preceding becomes a proof that

(DCB), (CBA) imply (DBA)

or equivalently (by II, 1),

(ABC), (BCD) imply (ABD)

which yields the other half of the theorem's conclusion.

Theorem B-6
Let A, B, C, and D be distinct points of a line. If (ABC) and (ACD), then (BCD) and (ABD).

This can be proved with similar techniques, as can the following more general result, which was listed as an axiom in Hilbert's original 1899 edition, but was subsequently found by E. H. Moore to depend on the other axioms.

Theorem B-7
Given any four distinct points on a line, they can be labeled A, B, C, and D in such a way that (ABC), (ABD), (BCD), and (ACD).

Definition B-8
For two distinct points A and B, the ray $\overrightarrow{AB}$ is the set consisting of the points of $\overline{AB}$ together with all points C such that (ABC). We say that A is the initial point of $\overrightarrow{AB}$ or that $\overrightarrow{AB}$ emanates from A.

Definition B-9
Let A, B, and C be three distinct points on a line a. We say that B and C lie on opposite sides of A if (BAC). We say that B and C lie on the same side of A if (BAC) is false, i.e., if either (ACB) or (ABC).

GROUP III: THE CONGRUENCE AXIOMS

III,1. If A, B are two points on a line a, and A' is a point on the same or on another line a' then it is always possible to find a point B' on a given side of the line a' through A' such that the segment $\overline{AB}$ is congruent or equal to the segment $\overline{A'B'}$. In symbols,

$$\overline{AB} \cong \overline{A'B'}$$

III,2. If a segment $\overline{A'B'}$ and a segment $\overline{A''B''}$ are congruent to the same segment $\overline{AB}$, then the segment $\overline{A'B'}$ is also congruent to the segment $\overline{A''B''}$, or briefly, if two segments are congruent to a third one they are congruent to each other.

Lemma B-10

Every segment is congruent to itself.

Proof. Using III,1 construct any segment $\overline{A'B'}$ congruent to $\overline{AB}$ and then apply III,2 to the congruences $\overline{AB} \cong \overline{A'B'}$ and $\overline{AB} \cong \overline{A'B'}$.

Theorem B-11

Congruence of segments is an equivalence relation.

Proof. Left to the reader. See Example vi of Section III-2 for the definition of equivalence relation.

III,3. On the line a let $\overline{AB}$ and $\overline{BC}$ be two segments which except for B have no point in common. Furthermore, on the same or on another line a' let $\overline{A'B'}$ and $\overline{B'C'}$ be two segments which except for B' also have no point in common. In that case, if

$$\overline{AB} \cong \overline{A'B'} \text{ and } \overline{BC} \cong \overline{B'C'}$$

then

$$\overline{AC} \cong \overline{A'C'}.$$

Definition B-12

Let [α be a plane and] h, k any two distinct rays emanating from O [in α] and lying in *distinct lines*. The pair of rays h, k is called an *angle* and is denoted by ∢(h,k) or by ∢(k,h). Rays h and k are the *sides* of the angle and point O is its *vertex*. Two angles with a vertex and one side in common and whose non-common sides form a

GROUP III: THE CONGRUENCE AXIOMS

line are called *supplementary*. An angle which is congruent to one of its supplementary angles is called a *right angle*.

The interior of an angle (or of a triangle) is defined in part d of Section III-3.

III,4. Let $\angle(h,k)$ be an angle [in a plane α] and a' a line [in a plane α'] and let a definite side of a' [in α'] be given. Let h' be a ray on the line a' that emanates from the point O'. Then there exists [in the plane α'] one and only one ray k' such that the angle $\angle(h,k)$ is congruent or equal to the angle $\angle(h',k')$ and at the same time all interior points of the angle $\angle(h',k')$ lie on the given side of a'. Symbolically, $\angle(h,k) \cong \angle(h',k')$. Every angle is congruent to itself, i.e., $\angle(h,k) \cong \angle(h,k)$ is always true.

Following standard practice, if A and B are points (other than O) on rays h and k, respectively, we shall usually write $\angle AOB$ or $\angle BOA$ rather than $\angle(h,k)$.

III,5. If for two triangles $\triangle ABC$ and $\triangle A'B'C'$ the congruences

$$\overline{AB} \cong \overline{A'B'}, \quad \overline{AC} \cong \overline{A'C'}, \quad \angle BAC \cong \angle B'A'C'$$

hold, then the congruence $\angle ABC \cong \angle A'B'C'$ is also satisfied.

Note: from an interchange of the letters B and C in III,5, it follows that under the same hypotheses $\angle ACB \cong \angle A'C'B'$ holds also.

From now on, we shall write AB = CD to indicate $\overline{AB} \cong \overline{CD}$. It is a consequence of Hilbert's postulates that real numbers (lengths) can be assigned to segments in such a way that segments are congruent if and only if they have the same length, and so that if the distance between two points A and B is defined to be the length of $\overline{AB}$, then the usual properties of a distance function hold. (We shall not go into the details here.) In this text, we shall use AB to denote the distance from A to B. Thus AB = CD indicates both congruence of segments and (equivalently) equality of distance.

Theorem B-13

The point B' whose existence is asserted in Axiom III,1 is unique (for a given side of A').

Proof. Suppose there were two (distinct) such points, B' and B" Choose any point C not on $\overleftrightarrow{A'B'}$, and apply III,5 to $\triangle A'B'C$ and $\triangle A'B''C$ to show that $\angle A'CB' \cong \angle A'CB''$. Then deduce a contradiction to the uniqueness asserted in III,4.

Theorem B-14

If, in $\triangle ABC$, $AB = AC$, then $\measuredangle ABC \cong \measuredangle ACB$.

Proof. Apply III,5 (and the last part of III,4) with $A' = A$, $B' = C$, $C' = B$.

Definition B-15

Two triangles will be called *congruent* if there exists a one-to-one correspondence between their vertices relative to which corresponding sides are congruent and corresponding angles are congruent. In other words, the triangles are congruent if we can label the vertices A, B, C in one triangle and A', B', C' in the other in such a way that the congruences

$$AB = A'B', \quad AC = A'C', \quad BC = B'C'$$

$$\measuredangle BAC \cong \measuredangle B'A'C', \quad \measuredangle ABC \cong \measuredangle A'B'C', \quad \measuredangle ACB \cong \measuredangle A'C'B'$$

all hold.

Theorem B-16 (Side-Angle-Side, or SAS)

If, in $\triangle ABC$ and $\triangle A'B'C'$, $AB = A'B'$, $AC = A'C'$, and $\measuredangle BAC \cong \measuredangle B'A'C'$, then $\triangle ABC = \triangle A'B'C'$.

Proof. We already know, from III,5 and the note following it, that $\measuredangle ABC \cong \measuredangle A'B'C'$ and $\measuredangle ACB \cong \measuredangle A'C'B'$ (Fig. B-4). It remains to prove $BC = B'C'$. Suppose on the contrary, that this were false. By III,1, choose D' on $\overrightarrow{B'C'}$ such that $B'D' = BC$. Applying III,5 to $\triangle ABC$ and $\triangle A'B'D'$ (where $\measuredangle ABC \cong \measuredangle A'B'D'$), we deduce that $\measuredangle BAC \cong \measuredangle B'A'D'$. Since $\measuredangle BAC \cong \measuredangle B'A'C'$ is given, we then have a contradiction to the uniqueness asserted in III,4, since $\overrightarrow{A'D'}$ and $\overrightarrow{A'C'}$ are distinct rays (why?).

From the preceding postulates and theorems, it is possible to deduce the transitivity of angle congruence, the remaining triangle

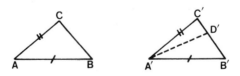

Figure B - 4.

GROUP III: THE CONGRUENCE AXIOMS

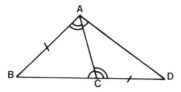

Figure B - 5.

congruence theorems (ASA, SSS, SAA), the congruence of vertical angles, the existence of right angles, the possibility of angle and segment bisection, and many other propositions of Euclid, Book I. The details appear in Hilbert [21]. As a final example we give Hilbert's proof of the exterior angle theorem.

Theorem B-17

An exterior angle of a triangle is greater than either of the interior angles not adjacent to it.

Proof. In Figure B-5, choose D on $\overline{BC}$ extended so that AB = CD. First, we shall show that ∡ACD ≇ ∡BAC. If ∡ACD ≅ ∡BAC were true then △BAC ≅ △DCA by Theorem B-16. But then ∡CAD would be congruent to the supplement of ∠ACD and hence congruent to the supplement of ∡BAC. By III,4, this in turn would imply that D lies on $\overleftrightarrow{BA}$, in violation of I,2. Thus, ∡ACD ≅ ∡BAC is impossible.

Next suppose ∡ACD < ∡BAC (Fig. B-6). Draw ray $\overrightarrow{AB'}$ on the B side of $\overleftrightarrow{AC}$ such that ∡B'AC ≅ ∡ACD. Then $\overrightarrow{AB'}$ passes through the interior of ∡BAC and so meets $\overline{BC}$ (Theorem III-17 in Section III-3) in a point, B'. But then the exterior angle ∡ACD of △B'AC is congruent to the interior angle ∡B'AC. This was shown to be impossible in the first part of the proof.

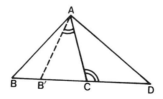

Figure B - 6.

The only possibility remaining is $\angle ACD > \angle BAC$. In like manner, we can prove that $\angle ACD > \angle ABC$.

GROUP IV: THE PARALLEL POSTULATE

Definition B-18

Two lines are said to be *parallel* if they [lie in the same plane and] do not intersect.

IV. (Euclid's Axiom) Let a be any line and A a point not on it. Then there is at most one line [in the plane determined by a and A,] that passes through A and does not intersect a.

GROUP V: THE CONTINUITY AXIOMS

V,1. (Axiom of Archimedes) If $\overline{AB}$ and $\overline{CD}$ are any segments then there exists a number n such that n segments $\overline{CD}$ constructed contiguously from A, along the ray from A to B, will pass beyond the point B.

V,2. (Linear Completeness Axiom) An extension of a set of points on a line with its order and congruence relations existing among the original elements as well as the fundamental properties of line order and congruence that follows from Axioms I-III, and from V,1 is impossible.

A more precise version of the Axiom of Archimedes appears as Theorem III-13 in Section III-3. The Linear Completeness Axiom may re restated as follows. In any model for the set of statements consisting of Axioms I,1-3 (except for the second part of I,3), II,1-3, III,1-3, V,1, and Theorems B-7 and B-13, it is impossible to adjoin any new points to a line in such a way that these statements are still true for the extended line. Roughly speaking, V,2 asserts that a line has no "holes" in it.

Many authors replace V,1 and V,2 by Dedekind's Postulate (see Section III-3), which is an equivalent assumption.

APPENDIX C

HYPERBOLIC FUNCTIONS

Definition C-1

For any real number x, we define

$$\sinh x = \frac{1}{2}(e^x - e^{-x}), \quad \cosh x = \frac{1}{2}(e^x + e^{-x})$$

$$\tanh x = \frac{\sinh x}{\cosh x} = \frac{e^x - e^{-x}}{e^x + e^{-x}}$$

$$\operatorname{sech} x = \frac{1}{\cosh x} = \frac{2}{e^x + e^{-x}}$$

The graphs of the first three of these functions are sketched in Figure C-1. The identities below follow readily from the definitions. As with the circular (ordinary trigonometric) functions, $\cosh^2 x$, $\sinh^2 x$, etc., denote $(\sinh x)^2$, $(\cosh x)^2$, etc.

$\cosh^2 x - \sinh^2 x = 1$

$1 - \tanh^2 x = \mathrm{sech}^2 x$

$\sinh(x \pm y) = \sinh x \cosh y \pm \cosh x \sinh y$

$\cosh(x \pm y) = \cosh x \cosh y \pm \sinh x \sinh y$

$\tanh(x \pm y) = \dfrac{\tanh x \pm \tanh y}{1 \pm \tanh x \tanh y}$

$\sinh 2x = 2 \sinh x \cosh x$

$\cosh 2x = \cosh^2 x + \sinh^2 x = 2 \cosh^2 x - 1$

$\dfrac{d}{dx} \sinh x = \cosh x, \quad \dfrac{d}{dx} \cosh x = \sinh x$

$\dfrac{d}{dx} \tanh x = \mathrm{sech}^2 x, \quad \dfrac{d}{dx} \mathrm{sech}\, x = -\mathrm{sech}\, x \tanh x$

From the power series for e^x and e^{-x},

$$e^x = 1 + x + \frac{x^2}{2!} + \frac{x^3}{3!} + \frac{x^4}{4!} + \cdots$$

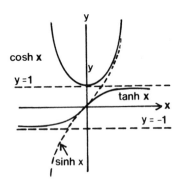

Figure C - 1.

HYPERBOLIC FUNCTIONS

$$e^{-x} = 1 - x + \frac{x^2}{2!} - \frac{x^3}{3!} + \frac{x^4}{4!} - \ldots$$

and the definitions above, we have

$$\sinh x = x + \frac{x^3}{3!} + \frac{x^5}{5!} + \frac{x^7}{7!} + \ldots$$

$$\cosh x = 1 + \frac{x^2}{2!} + \frac{x^4}{4!} + \frac{x^6}{6!} + \ldots$$

$$\tanh x = x - \frac{x^3}{3} + \frac{2x^5}{15} - \frac{17x^7}{315} + \ldots, \quad |x| < \frac{\pi}{2}$$

Except for the differences in signs, these resemble the series for the circular functions:

$$\sin x = x - \frac{x^3}{3!} + \frac{x^5}{5!} - \frac{x^7}{7!} + \ldots$$

$$\cos x = 1 - \frac{x^2}{2!} + \frac{x^4}{4!} - \frac{x^6}{6!} + \ldots$$

$$\tan x = x + \frac{x^3}{3!} + \frac{2x^5}{15} + \frac{17x^7}{315} + \ldots, \quad |x| < \frac{\pi}{2}$$

This resemblance is more than a coincidence, as is shown in Section VI-6.

APPENDIX D

VECTOR GEOMETRY AND ANALYSIS

Intuitively speaking, a vector is an entity that has both magnitude and direction, and so is often represented as a directed line segment or "arrow." Arrows with the same length and direction are considered to be equivalent and represent the same vector. Accordingly, many authors define a vector to be an equivalence class of directed line segments, i.e., a set consisting of an arrow together with all arrows having the same length and direction as it.

In E^3, if a vector **V** is represented by an arrow with initial point (x,y,z) and terminal point (x',y',z'), then the coordinate differences,

$$\Delta x = x' - x, \quad \Delta y = y' - y, \quad \Delta z = z' - z$$

are called the *components* of **V**. It is easy to see that equivalent arrows have the same components, so the choice of representing arrow is immaterial.

Throughout this appendix, we treat vectors in 3-dimensional Euclidean space, E^3. However, with the exception of the cross product, which is defined in three dimensions only, analogous definitions and results pertain in two dimensions: simply delete the third coordinate of every point and the third component of every vector.

For each point, we can associate the vector represented by the arrow from the origin to that point. Conversely, each vector has a representing arrow emanating from the origin and may be associated with the terminal point of that arrow. The coordinates of the point are exactly the components of the vector. Since points and vectors are thus in one-to-one correspondence, we shall use ordered triples to label vectors as well as points, e.g., $V = (a,b,c)$. The context will usually make clear which is intended, but in many instances we shall want to think of the same object as a point at certain times and a vector at others.

Definition D-1

If $V_1 = (a_1, a_2, a_3)$ and $V_2 = (b_1, b_2, b_3)$ are vectors and k is any real number, we define

$$V_1 + V_2 = (a_1 + b_1, a_2 + b_2, a_3 + b_3)$$

$$kV_1 = (ka_1, ka_2, ka_3)$$

Geometrically, vector addition is accomplished by choosing representatives for the vectors for which the terminal point of the first vector coincides with the initial point of the second. The sum is then the arrow from the initial point of the first to the terminal point of the second. This is illustrated in Figure D-1 for the 2-dimensional case.

Vector addition is obviously commutative and satisfies

$$(k_1 + k_2)V = k_1 V + k_2 V$$

$$k(V_1 + V_2) = kV_1 + kV_2$$

We define $V - W$ to mean $V + (-1)W$. The geometric significance of

VECTOR GEOMETRY AND ANALYSIS

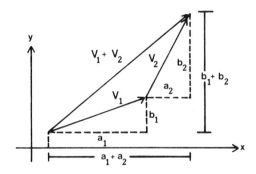

Figure D - 1.

vector subtraction is shown in Figure D-2. The vector whose components are all zero is denoted **O**.

Definition D-2

The length of a vector $\mathbf{V} = (a,b,c)$ is the non-negative number $\|\mathbf{V}\|$ given by

$$\|\mathbf{V}\| = (a^2 + b^2 + c^2)^{1/2}$$

It is the distance between the initial and terminal points of any arrow representing **V**.

The length $\|\mathbf{V}\|$ is zero if and only if $\mathbf{V} = \mathbf{O}$. From the geometric interpretation of vector addition (cf. Fig. D-1), we have the *triangle inequality*,

$$\|\mathbf{V} + \mathbf{W}\| \leq \|\mathbf{V}\| + \|\mathbf{W}\|$$

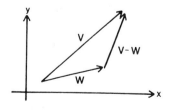

Figure D - 2.

for all **V** and **W**. For any vector **V** and real number k,

$$\|k\mathbf{V}\| = |k|\, \|\mathbf{V}\|$$

As a consequence, if $\mathbf{V} \neq \mathbf{O}$, the vector $(1/\|\mathbf{V}\|)\mathbf{V}$, abbreviated $\mathbf{V}/\|\mathbf{V}\|$, is a *unit vector*, i.e., a vector of length one, having the same direction as **V**.

Definition D-3
  The *scalar product* or *dot product* of two vectors $\mathbf{V} = (a_1, b_1, c_1)$ and $\mathbf{W} = (a_2, b_2, c_2)$ is the real number $\mathbf{V} \cdot \mathbf{W}$ defined by

$$\mathbf{V} \cdot \mathbf{W} = a_1 a_2 + b_1 b_2 + c_1 c_2$$

Scalar multiplication is commutative and distributive with respect to vector addition. For any V, $\mathbf{V} \cdot \mathbf{V} = \|\mathbf{V}\|^2$.

Theorem D-4
  If $\mathbf{V} = (a_1, b_1, c_1)$ and $\mathbf{W} = (a_2, b_2, c_2)$ then

$$\mathbf{V} \cdot \mathbf{W} = \|\mathbf{V}\|\, \|\mathbf{W}\| \cos \theta$$

where θ is the angle between **V** and **W**.

Proof. We give the proof in two dimensions, the proof for three being entirely similar. Applying the Law of Cosines to $\triangle OPQ$ in Figure D-3, we obtain

$$(a_2 - a_1)^2 + (b_2 - b_1)^2 = (PQ)^2 = \|\mathbf{V}\|^2 + \|\mathbf{W}\|^2 - 2\|\mathbf{V}\|\,\|\mathbf{W}\|\cos\theta$$

$$= a_1^2 + b_1^2 + a_2^2 + b_2^2 - 2\|\mathbf{V}\|\,\|\mathbf{W}\|\cos\theta$$

or
$$2\|\mathbf{V}\|\,\|\mathbf{W}\|\cos\theta = 2(a_1 a_2 + b_1 b_2) = 2(\mathbf{V}\cdot\mathbf{W})$$

Corollary D-5
  Two vectors **V** and **W** are *orthogonal* (i.e., perpendicular when drawn emanating from the same point) if and only if $\mathbf{V}\cdot\mathbf{W} = 0$. (The zero vector, **O**, is, by convention, orthogonal to every vector.)

VECTOR GEOMETRY AND ANALYSIS

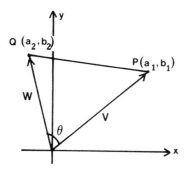

Figure D - 3.

Definition D-6

The *vector product* or *cross product* of two vectors $\mathbf{V} = (a_1, b_1, c_1)$ and $\mathbf{W} = (a_2, b_2, c_2)$ is the vector $\mathbf{V} \times \mathbf{W}$ given by

$$\mathbf{V} \times \mathbf{W} = (b_1 c_2 - b_2 c_1,\ c_1 a_2 - c_2 a_1,\ a_1 b_2 - a_2 b_1)$$

The cross product can be computed by formally evaluating the determinant of the 3×3 matrix

$$\begin{pmatrix} \mathbf{i} & \mathbf{j} & \mathbf{k} \\ a_1 & b_1 & c_1 \\ a_2 & b_2 & c_2 \end{pmatrix}$$

where $\mathbf{i} = (1,0,0)$, $\mathbf{j} = (0,1,0)$, and $\mathbf{k} = (0,0,1)$. Thus, expanding this determinant along the first row, we have

$$\mathbf{V} \times \mathbf{W} = \begin{vmatrix} b_1 & c_1 \\ b_2 & c_2 \end{vmatrix} \mathbf{i} - \begin{vmatrix} a_1 & c_1 \\ a_2 & c_2 \end{vmatrix} \mathbf{j} + \begin{vmatrix} a_1 & b_1 \\ a_2 & b_2 \end{vmatrix} \mathbf{k}$$

$$= (b_1 c_2 - b_2 c_1,\ c_1 a_2 - c_2 a_1,\ a_1 b_2 - a_2 b_1)$$

The following identities can be verified by computation with components.

$$\mathbf{V} \times \mathbf{W} = -(\mathbf{W} \times \mathbf{V})$$
$$\mathbf{V} \times \mathbf{V} = \mathbf{O}$$
$$\mathbf{V} \times (\mathbf{W}_1 + \mathbf{W}_2) = (\mathbf{V} \times \mathbf{W}_1) + (\mathbf{V} \times \mathbf{W}_2)$$
$$k_1 \mathbf{V}_1 \times k_2 \mathbf{V}_2 = k_1 k_2 (\mathbf{V}_1 \times \mathbf{V}_2)$$
$$\mathbf{U} \cdot (\mathbf{V} \times \mathbf{W}) = \mathbf{V} \cdot (\mathbf{W} \times \mathbf{U}) = \mathbf{W} \cdot (\mathbf{U} \times \mathbf{V})$$

The quantity $\mathbf{U} \cdot (\mathbf{V} \times \mathbf{W})$ is called the *triple scalar product* of $\mathbf{U}$, $\mathbf{V}$ and $\mathbf{W}$, and the last property above is often called the *cyclic property* of the triple scalar product. We usually omit the parentheses in $\mathbf{U} \cdot (\mathbf{V} \times \mathbf{W})$ since there is only one possible way of interpreting $\mathbf{U} \cdot \mathbf{V} \times \mathbf{W} - (\mathbf{U} \cdot \mathbf{V}) \times \mathbf{W}$ makes no sense, since we cannot cross a number and a vector.

If $\mathbf{V}_i = (a_i, b_i, c_i)$, for $i = 1, 2, 3$, then $\mathbf{V}_1 \cdot \mathbf{V}_2 \times \mathbf{V}_3$ is the determinant

$$\mathbf{V}_1 \cdot \mathbf{V}_2 \times \mathbf{V}_3 = \begin{vmatrix} a_1 & b_1 & c_1 \\ a_2 & b_2 & c_2 \\ a_3 & b_3 & c_3 \end{vmatrix}$$

Theorem D-7

Let $\mathbf{V}$ and $\mathbf{W}$ be non-zero vectors and let θ be the angle between $\mathbf{V}$ and $\mathbf{W}$ ($0 \leq \theta \leq \pi$). Then (a) $\mathbf{V} \times \mathbf{W}$ is othogonal to both $\mathbf{V}$ and $\mathbf{W}$, (b) $\|\mathbf{V} \times \mathbf{W}\| = \|\mathbf{V}\| \, \|\mathbf{W}\| \sin \theta$.

Proof. (a) can be verified by computing $\mathbf{V} \cdot \mathbf{V} \times \mathbf{W}$ and $\mathbf{W} \cdot \mathbf{V} \times \mathbf{W}$ in components, or by using the fact that the determinant of a matrix with two equal rows is zero. For (b),

$$\|\mathbf{V}\|^2 \|\mathbf{W}\|^2 \sin^2 \theta = \|\mathbf{V}\|^2 \|\mathbf{W}\|^2 (1 - \cos^2 \theta) = \|\mathbf{V}\|^2 \|\mathbf{W}\|^2 - (\mathbf{V} \cdot \mathbf{W})^2$$

If the latter is computed in components, it will be found to agree with $\|\mathbf{V} \times \mathbf{W}\|^2$ expressed in components.

The length of the cross product, $\|\mathbf{V} \times \mathbf{W}\|$, can be interpreted geometrically as the area of a parallelogram in which two adjacent sides are arrows representing $\mathbf{V}$ and $\mathbf{W}$. In Figure D-4, area = base × height = $\|\mathbf{V}\| \, \|\mathbf{W}\| \sin \theta$. If $\mathbf{V} \times \mathbf{W} \neq \mathbf{O}$, the direction of $\mathbf{V} \times \mathbf{W}$ is such that $\mathbf{V}$, $\mathbf{W}$, and $\mathbf{V} \times \mathbf{W}$ form a "right-handed system": if you extend the index and middle fingers and thumb of your right hand so that (a) your index finger points in the direction of $\mathbf{V}$, (b) your middle finger points in the direction of $\mathbf{W}$, and (c) your thumb is

VECTOR GEOMETRY AND ANALYSIS

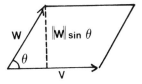

Figure D - 4.

perpendicular to both fingers, then your thumb will point in the direction of **V** × **W**.

Occasionally, we deal with *vector-valued functions* of a real variable t, for example,

$$\mathbf{V}(t) = (a(t), b(t), c(t))$$

where a, b, and c are differentiable real-valued functions. The derivative is defined componentwise:

$$\frac{d\mathbf{V}}{dt} = \left(\frac{da}{dt}, \frac{db}{dt}, \frac{dc}{dt}\right)$$

The following are immediate consequences of the sum, product, and quotient rules for real-valued functions.

$$\frac{d}{dt}(\mathbf{V} + \mathbf{W}) = \frac{d\mathbf{V}}{dt} + \frac{d\mathbf{W}}{dt}$$

$$\frac{d}{dt}(\mathbf{V} \cdot \mathbf{W}) = \mathbf{V} \cdot \frac{d\mathbf{W}}{dt} + \frac{d\mathbf{V}}{dt} \cdot \mathbf{W}$$

$$\frac{d}{dt}(f\mathbf{V}) = f\frac{d\mathbf{V}}{dt} + \frac{df}{dt}\mathbf{V}$$

$$\frac{d}{dt}\left(\frac{1}{f}\mathbf{V}\right) = \frac{f\frac{d\mathbf{V}}{dt} - \frac{df}{dt}\mathbf{V}}{f^2}$$

$$\frac{d}{dt}(\mathbf{V} \times \mathbf{W}) = \mathbf{V} \times \frac{d\mathbf{W}}{dt} + \frac{d\mathbf{V}}{dt} \times \mathbf{W}$$

for **V** and **W** vector functions and f a real-valued function. We shall interchangeably write f**V** or **V**f. If **V** = **V**(s) and s = f(t), then the identity

$$\frac{d\mathbf{V}}{dt} = \frac{d\mathbf{V}}{ds}\frac{ds}{dt}$$

follows from the ordinary chain rule for real-valued functions. If **W** is a vector function of two real variables. **W** = **W**(u,v) and u and v in turn are functions of a variable s, then

$$\frac{d\mathbf{W}}{ds} = \frac{\partial \mathbf{W}}{\partial u}\frac{du}{ds} + \frac{\partial \mathbf{W}}{\partial v}\frac{dv}{ds}$$

(All functions are assumed differentiable.)

BIBLIOGRAPHY

1. Abbott, Edwin A., *Flatland*, 6th ed., Dover Publications, New York, 1952
2. Ball, W. W. R., *A Short Account of the History of Mathematics*, Dover Publications, New York, 1960
3. Bell, E. T., *Men of Mathematics*, Simon & Schuster, New York, 1965
4. Birkhoff, G. D., "A Set of Postulates for Plane Geometry," *Annals of Mathematics*, 33 (1932)
5. Bonola, R., *Non-Euclidean Geometry*, tr. by H. S. Carslaw, Dover Publications, New York, 1955
6. Borsuk, Karol, and W. Szmielew, *Foundations of Geometry*, tr. by E. Marquit, Interscience Publishers, New York, 1960
7. Boyer, C. B., *A History of Mathematics*, John Wiley & Sons, New York, 1968
8. Chace, A. B., et. al., *The Rhind Mathematical Papyrus*, 2 vols., Mathematical Association of America, 1927-1929
9. Cole, R. H., *Theory of Ordinary Differential Equations*, Appleton-Century-Crofts, New York, 1968
10. Coolidge, J. L., *A History of Geometrical Method*, Dover Publications, New York, 1963
11. Cough, A. H., *Plutarch's Lives*, Dryden's translation, Little, Brown & Co., Boston, 1899
12. Court, N. A., "The Problem of Apollonius," *The Mathematics Teacher*, 54 (1961), 444-452

13. Eves, H., *A Survey of Geometry,* Vol. 1, Allyn and Bacon, Inc., Boston, 1963
14. Gauss, C. F., *General Investigations of Curved Surfaces*, translated by J. C. Morehead and A. M. Hiltebeitel, repr., Raven Press, New York, 1965
15. Gillings, Richard J., *Mathematics in the Time of the Pharaohs,* MIT Press, Cambridge, 1972
16. Greenberg, M. J., *Euclidean and Non-Euclidean Geometries,* 2nd ed., W. H. Freeman and Co., San Francisco, 1980
17. Hall, Tord, *Carl Friedrich Gauss, A Biography,* tr. by A. Froderberg, MIT Press, Cambridge, 1970
18. Heath, T. L., *The Thirteen Books of Euclid's Elements,* 2nd ed., 3 vols., Dover Publications, New York, 1956
19. Heath, T. L., *The Works of Archimedes,* repr. of 1897 ed., Dover Publications, Inc., New York
20. Hilbert, D. and S. Cohn-Vossen, *Geometry and the Imagination,* Chelsea Publishing Co., New York, 1952
21. Hilbert, David, *Foundations of Geometry,* 2nd ed., tr. by Leo Unger from the 10th German ed., Open Court Publishing Co., La Salle, Ill., 1971
22. Hoyle, Fred, *Astronomy,* Rathbone Books Ltd., London, England, 1962
23. Kay, David C., *College Geometry,* Holt, Rinehart and Winston, Inc., New York, 1969
24. Kennedy, H. C., "The Origins of Modern Axiomatics: Pasch to Peano," *American Mathematical Monthly,* 79 (1972), 133-136
25. Kline, Morris, "Projective Geometry," in *The World of Mathematics,* Vol. 1, Simon & Schuster, New York, 1956
26. Kline, Morris, *Mathematical Thought from Ancient to Modern Times,* Oxford University Press, New York, 1972
27. Kulczycki, S., *Non-Euclidean Geometry,* tr. by S. Knapowski, Pergamon Press, Ltd., Oxford, 1961
28. Legendre, A. M., *Elements of Geometry and Trigonometry,* tr. by D. Brewster, A. S. Barnes & Co., Hartford, 1838
29. Martin G., *The Foundations of Geometry and the Non-Euclidean Plane,* Intext Educational Publishers, New York, 1972
30. McCoy, Neal H., *Fundamentals of Abstract Algebra,* Allyn and Bacon, Inc., Boston, 1972
31. Meschowski, H., *Noneuclidean Geometry,* tr. by A. Shenitzer, Academic Press, New York, 1964

BIBLIOGRAPHY

32. Neugebauer, O., *The Exact Sciences in Antiquity,* 2nd ed., Brown University Press, Providence, 1957
33. Neugebauer, O., and A. Sachs, "Mathematical Cuneiform Texts," *American Oriental Series* (29), Yale University Press, New Haven, 1945
34. Ogilvy, C. Stanley, *Excursions in Geometry,* Oxford University Press, New York, 1969
35. Pogorelov, A. V., *Lectures on the Foundations of Geometry,* tr. by Leo F. Boron, P. Noordhoff Ltd., Groningen, 1966
36. Poincaré, H., *The Foundations of Science,* tr. by G. B. Halsted, The Science Press, Lancaster, Pa., 1913
37. Pontryagin, L., *Ordinary Differential Equations,* tr. by W. Counts, Addison-Wesley Publishing Co., Reading, Ma., 1962
38. Rucker, Rudolf v. B., *Geometry, Relativity and the Fourth Dimension,* Dover Publications, New York, 1977
39. Saccheri, G., *Euclides ab Omni Naevo Vindicatus,* tr. by G. B. Halsted, repr., Chelsea Publishing Co., New York, 1970
40. Sawyer, W. W., *A Concrete Approach to Abstract Algebra,* W. H. Freeman and Co., San Francisco, 1959
41. *School Mathematics Study Group: Geometry,* Yale University Press, New Haven, 1961
42. Shirakov, P. A., *A Sketch of the Fundamentals of Lobachevskian Geometry,* tr. by Leo Boron, P. Noordhoff Ltd., Groningen, 1964
43. Smith, D. E., *History of Mathematics,* 2 vols., Dover Publications, New York, 1958
44. Sommerville, D. M. Y., *The Elements of Non-Euclidean Geometry,* Dover Publications, New York, 1958
45. Stackel, Paul, "Wolfgang und Johann Bolyai," *Geometriche Untersuchungen,* Vol. I, Leipzig and Berlin, 1913
46. Stewart, Ian, "Gauss," *Scientific American,* 237 (1977), 122-131
47. Struve, W. W., "Mathematisch Papyrus des Museums in Moskau," *Quellen und Studien zur Geschichte der Mathematik,* Ser. A, Vol. 1, Berlin, 1930
48. Tuller, Anita, *A Modern Introduction to Geometries,* Van Nostrand Reinhold Co., New York, 1967
49. Von Fritz, K., "The Discovery of Incommensurables by Hippasus of Metapontum," *Annals of Mathematics,* 46 (1945), 242-264

50. Waerden, B. L. van der, *Science Awakening,* Oxford University Press, New York, 1961
51. White, A. J., *Real Analysis: An Introduction*, Addison-Wesley Publishing Co., London, 1968
52. Wilder, R. L., "The Axiomatic Method," in *The World of Mathematics,* Vol. 3, Simon & Schuster, New York, 1956
53. Wolfe, Harold E., *Introduction to Non-Euclidean Geometry,* Holt, Rinehart and Winston, Inc., New York, 1945
54. Wylie, C. R., Jr., *Foundations of Geometry,* McGraw-Hill Book Co., New York, 1964

INDEX

A

Absolute Geometry, 130-141, 160
Absolute Lengths, 141-143, 148, 203, 226
Ahmes, 29-33, 35, 38, 41-43
Angle Criterion, 171, 214
Angles:
    dihedral, 207
    in H^2, 265-269
    of parallelism, 169, 176 180, 185, 230
    trisection of, 53, 72
Angle sum:
    in a polygon, 139, 149
    in a triangle, 131, 132, 134, 137
Aphelion, 84
Apollonius, 60, 74-80
    astronomy of, 83
    the problem of, 77
Archimedes, 56, 60, 65-74
    Axiom of, 55, 111, 123, 145
    on floating bodies, 71

(Archimedes)
    screw of, 73
    spiral of, 70, 72
Archytas, 49, 51, 81
Area:
    in Lobachevskian geometry, 150, 151, 238, 244, 280
    in spherical geometry, 159
Aristarchus, 82, 83
Athenian School, 53
Axioms, 92
    categoricity of, 106
    completeness of, 106
    consistency of, 105-108
    independence of, 105-108

B

Babylonians, 2-25, 55
Beltrami, 105, 224 (*see also* Klein-Beltrami model and Pseudosphere)
Betweenness, 115, 196, 255
Bisector of the strip, 172
Bolyai, J., 160-162, 164
Bolyai, W., 154-155

C

Categoricity, 106
Circles:
    area of, 11, 17, 32, 59, 65, 244, 245, 287
    circumference of, 11, 17, 244
    great, 113-115
    in H^2, 259
    inside and outside of, 110
    in Lobachevskian plane, 114, 244, 245
    squaring of, 53
    theorems on, 62
Completeness, 106
Congruence of triangles:
    ASA, 46, 61
    in Lobachevskian geometry, 142
    SAA, 61
    SAS, 61, 109, 113, 267
    SSS, 61
Conic Sections, 57, 74-76
Consistency, 105-108
Continuity, 109, 145
    Dedekind's Postulate of, 109-112
Contrapositive, 126, 128, 129, 140
Converse, 126, 128, 129
Correspondence of Points, 173, 193, 219, 227
Cosmology, 78, 80-90
Cuneiform, 3
Curvature, 109, 285, 286

D

Dedekind, 55, 59
    Postulate of, 109-112

Defect:
    and parallax, 280
    of a polygon, 139, 149
    of a triangle, 132-134, 157, 236
Delian problem (*see* Duplication of the cube)
Duplication of the cube, 50, 51, 57

E

Ecliptic, 85
Egyptians, 25-44
Elliptic Geometry, 277, 286
Epicycle, 88
Equidistant curves, 194, 196, 197
    in H^2, 275
Equidistant surface, 227
Equivalence relation, 107
Eratosthenes, 89
Euclid, 59-64
    Elements of, 54, 55, 60, 61, 91, 95, 126-130
    and foundations of geometry, 108-123
    other works of, 60
Eudoxus, 54-56, 62, 81, 82
Exhaustion, method of, 54, 56, 57, 59, 65, 67
Exterior Angle Theorem, 113, 117, 119, 127
    for ideal triangles, 183

G

Gauss, 155-159
    letters of, 154, 156, 158, 159, 161, 164
Geodesics, 225, 288

INDEX

H

H^2:
    definition of, 253
    distances in, 257
Half-plane, 116
Half-space, 123
Heron's formula, 70
Hieratic, 25-27
Hieroglyphics, 25, 27
Hilbert, 109, 110, 225
Hipparchus, 5, 83, 88, 89
Hippasus, 48, 49
Horocycles, 194
    concentric, 194, 199, 202
    in H^2, 272
Horosphere, 220
Hyperbolic geometry, 286

I

Incommensurables, 49, 54-56, 58, 59, 63, 201
Independence, 105-108
Infinitude of the line, 113, 115, 145
Inverse, 126
Isomorphism, 100, 101, 106

K

Kant, 157, 158
Klein-Beltrami Model, 136, 172, 182, 189, 218, 278

L

Lagrange, 151
Lambert, 147-151
    quadrilateral, 147-149, 236
Law of cosines, 235, 242-244
Law of sines, 160, 235
Legendre, 152, 180

Light Cone, 253
Limiting curves (*see* Horocycles)
Limiting surfaces (*see* Horosphere)
Lines:
    divergent, 172, 186-188, 275
    in H^2, 254, 262-264
    intersecting, 190, 256, 274
    parallel, 126, 171, 174, 213, 237
    sheaves of, 190
Lobachevsky, 162-165

M

Magnitudes, 54
Menaechmus, 57, 60, 74
Mesopotamia, 2, 3 (*see also* Babylonians)
    mathematics of, 2-25
Model, 99

N

Neugebauer, 7, 9, 15
Number systems:
    Egyptian, 25-30, 38-43
    positional notation, 3, 4
    sexagesimal, 4, 5, 13, 22-24, 89

P

Pappus, 45, 78
    Theorem of, 79, 80
Parabolic geometry, 286
Parallax of stars, 281-283
Parallel lines:
    Bolyai's construction of, 237
    Euclid's definition of, 126

(Parallel lines)
    in Lobachevskian plane, 171
    in Lobachevskian space, 213
Parallelism:
    angle of, 169, 179, 230
    of a line to a plane, 214
    of lines, 126, 171, 174, 213, 237
    of planes, 215
    of rays, 167, 185
Parallel Postulate:
    of Euclid, 126-128, 137, 139
    of Lobachevsky, 135, 137, 167
Pasch's Axiom, 115-117, 121, 122
    for ideal triangles, 182
Perihelion, 84
Perpendicularity:
    of a line and plane, 206
    of two planes, 208
Pi, approximations to, 11, 17, 33, 34, 41, 66, 89
Π:
    definition of, 169
    formula for, 230
    properties of, 176-180, 185
Plato, 53
Playfair's Axiom, 128, 129, 135, 139
Plimpton 322, 12-15, 17
Plutarch, 53
Poincaré, 136, 288
    Disk Model, 136, 172, 182, 190
    Upper Half-Plane, 137, 182, 190

Points:
    correspondence of, 173, 193, 219, 227
    ideal, 181, 185, 214, 216
    ultra-ideal, 188, 226
Pole, 77, 277
Polar, 77, 277
Postulate (*see* Axioms)
Proclus, 45, 127
Projection of a line onto a plane, 210
Projective plane, 115
Proportion, 54-56, 62, 152
Pseudosphere, 224
Pyramid:
    Great, 37, 47
    volume of a, 12, 34, 35, 43
Ptolemaic Theory, 83, 89
Ptolemy, 5, 83, 89
Pythagoras, 14, 48, 49
Pythagorean School, 48, 49, 53, 81
Pythagorean Theorem, 8, 13, 37, 49-51, 61, 130
    in Lobachevskian geometry, 232

R

Retrograde motion, 82
Rhind Mathematical Papyrus (RMP), 29-33, 35, 38-42
Riemann, 113

S

Saccheri, 143-146
    quadrilaterals, 143, 146, 147, 149-151, 238
Seleucid period, 3-5

INDEX

Semicircles, angles inscribed in, 46, 140, 153
Separation Axiom, 116, 117, 119, 122
Space Constant, 158, 203, 282
Sphere:
    area of, 36, 37, 287
    volume of, 68, 72, 286
Spherical geometry, 114, 143, 148, 277, 286
    area in, 151, 287
Sumerians, 2, 3
Superposition, 109

T

Thales, 46-48
Triangles:
    area of, 150, 151, 159, 238, 244, 280
    congruence of (*see* Congruence of triangles)

(Triangles)
    ideal, 182
    identities for, 232-235, 241
    perpendicular bisectors of, 191
    similar, 141
Trisection of an angle, 53, 72

U

Undefined terms, 95, 99

V

Vectors, 248

W

Weierstrass coordinates:
    of a point, 252
    of a line, 253